Concepts in
PLANT PATHOLOGY

The Authors

Dr. Abhishek Singh is working as an Assistant Professor (Plant Pathology) at Career Point University Kota, Rajasthan. He did his M.Sc. (Ag.) Plant Pathology and Ph.D. in Plant Pathology from Acharya Narendra Deva University of Agriculture and Technology, Kumarganj, Ayodhya, Uttar Pradesh. He has published 11 research paper in Journals of repute, 14 book chapters and several popular/ semi scientific articles.

Amit Kumar Singh obtained his M.Sc. (Ag) Agronomy from Acharaya Narendra Deva University of Agriculture and Technology, Kumarganj, Ayodhya (U.P.). He has qualified ARS (ICAR-NET) conducted by Agricultural Scientists Recruitment Board, New Delhi. Currently working as Block Technical Manager (BTM), ATMA, Paliganj, Patna. He worked as Senior Research Fellow (SRF) at ICAR-Research Complex for Eastern Region, Patna in research project sanctioned by Bill Gates Foundation, United States. He has vast experience in popularization of Agricultural farming techniques in the farmer field. He has published three books and number of research papers in national and international journals of repute and also organized/ participated various training programmes, seminars, symposia and workshops.

Concepts in
PLANT PATHOLOGY

– Authors –

Dr. Abhishek Singh

Assistant Professor (Plant Pathology)
Career Point University Kota, Rajasthan

Amit Kumar Singh

Block Technical Manager (BTM)
ATMA, Paliganj, Patna

2024

Daya Publishing House®
A Division of
Astral International Pvt. Ltd.
New Delhi – 110 002

ISBN: 978-93-5919-465-3

Published by : **Daya Publishing House®**
A Division of
Astral International Pvt. Ltd.
– ISO 9001:2015 Certified Company –
4736/23, Ground Floor, Ansari Road,
Darya Ganj, New Delhi-110 002
Ph. 011-43549197, 23278134
E-mail: info@astralint.com
Website: www.astralint.com

Digitally Printed at : **Replika Press Pvt. Ltd.**

Preface

Plant pathology, a crucial discipline within agricultural science, delves into the intricate world of plant diseases, their origins, mechanisms, and the development of strategies to mitigate their impact. As an indispensable field, plant pathology plays a pivotal role in sustaining global food security, economic stability, and environmental health. By investigating the interactions between pathogens and plants, understanding disease cycles, and developing innovative solutions, plant pathologists contribute significantly to the resilience and productivity of agricultural systems. In the face of evolving challenges such as climate change, globalization, and emerging pathogens, the insights garnered from plant pathology research are instrumental in fostering sustainable practices and ensuring the well-being of both cultivated crops and natural ecosystems. This field's multifaceted nature underscores its importance in addressing complex issues at the intersection of agriculture, ecology, and human well-being. Plant pathology is crucial for safeguarding global food security, promoting sustainable agriculture, and protecting natural ecosystems.

The knowledge and expertise of plant pathologists contribute to the development of resilient and productive agricultural systems, benefiting both human societies and the environment. Plant diseases have a profound economic impact on agriculture, influencing both crop quantity and quality. The financial consequences stem from reduced yields, increased production costs, and the need for extensive disease management measures. Yield losses result in decreased revenue for farmers and threaten global food security by limiting the availability of essential crops. Additionally, the expenses associated with the purchase of pesticides, fungicides, and other control measures escalate, contributing to the overall economic burden. The susceptibility of certain crops to specific diseases further intensifies the economic impact, affecting entire agricultural sectors and regional economies. Beyond direct agricultural consequences, plant diseases can disrupt supply chains, leading to higher commodity prices and financial instability in related industries. As agriculture is a cornerstone of many economies, the economic repercussions of plant diseases extend far beyond the farm, emphasizing the urgent need for effective disease management strategies and research in plant pathology to mitigate these financial challenges. Plant pathologists play a key role in educating farmers, agricultural professionals, and the general public about plant diseases and their management. Extension services provide valuable information and recommendations to help prevent and control plant diseases at the community level. Some plant diseases can have indirect effects on human health by contaminating food crops with harmful toxins or reducing the nutritional value of crops. Plant pathologists work to identify and mitigate such risks to ensure the safety and quality of the food supply.

This book has been designed to under-graduate students of agriculture in one compact volume of twelve chapters. These chapters include introduction to plant pathology, history of plant pathology, Indian plant pathologist, importance of plant diseases, classification of plant diseases, pathogenesis-process of disease development, phytopathogenic bacteria and mollicutes, viruses and plant diseases, fungi and plant diseases, nematodes and plant diseases, phanerogamic plant parasites and management of plant diseases. Related glossary and references have been provided at the end for the ready reference of readers.

It hoped that this book will be immense value to undergraduate students of agriculture, forestry, horticulture and allied subjects, teachers, researchers, extension workers, NGOs, other stakeholders involved in farm advisory services such as Agri-input dealers, Agripreneurs, Kisan Call Centers and extension functionaries working at grass roots level. It should be useful not only to students and professionals in plant pathology, but to those in allied fields such as plant science and botany, who are increasingly interested in plant diseases. This is a useful source of information for farmers, trainers, and other interested persons to improve not only the agriculture but also the livelihood of the farming community.

We are grateful to all those persons as well as various books, manuals, periodicals, newsletters, magazines, journals *etc.* that helped in the preparation of this book. In spite of the best efforts, it is possible that some errors may have occurred into the compilation and editing of the book. Further queries, constructive suggestions and criticisms for the further improvement of the book are always welcome and shall be thankfully acknowledged.

Abhishek Singh

Amit Kumar Singh

Contents

Chapter 1

Introduction to Plant Pathology

Introduction: An Overview

It seems like you'd like an introduction to plant pathology, which is the study of plant diseases. Plant pathology is a crucial field of science that focuses on understanding, preventing, and managing diseases that affect plants. These diseases can have significant economic and ecological impacts, as they can reduce crop yields, affect the quality of harvested produce, and even lead to the death of plants.

Plant diseases are pathological conditions that affect the health and vitality of plants, including crops, ornamental plants, trees, and other vegetation. These diseases can be caused by various agents, including pathogens (such as fungi, bacteria, viruses, and nematodes), environmental factors, and abiotic stressors. Plant diseases can have significant economic and ecological impacts, as they can reduce crop yields, damage landscapes, and disrupt ecosystems.

Basic Overview of Plant Pathology

- ☆ **Definition**: Plant pathology is the scientific study of plant diseases caused by pathogens, such as fungi, bacteria, viruses, nematodes, and parasitic plants, as well as abiotic factors like environmental stress.
- ☆ **Importance**: Plant diseases can lead to substantial losses in agriculture, forestry, and horticulture, affecting food security and the economy. Understanding and managing plant diseases are essential for sustainable agriculture.

Types of Plant Pathogens

- **Fungi:** Fungal pathogens are responsible for a significant portion of plant diseases. They can infect various plant parts, causing symptoms like wilting, rotting, and leaf spots.
- **Bacteria:** Bacterial pathogens can cause diseases in plants, resulting in symptoms like leaf blights and crown galls.
- **Viruses:** Plant viruses can disrupt normal plant growth and development, leading to symptoms like mosaic patterns and stunted growth.
- **Nematodes:** Plant-parasitic nematodes are microscopic worms that can damage plant roots and affect nutrient uptake.
- **Parasitic Plants:** Some plants, like dodder and broomrape, are parasitic and can infect and harm other plants.
- **Abiotic Factors:** Environmental stressors, such as drought, excessive moisture, temperature extremes, and pollution, can weaken plants and make them more susceptible to diseases.

Disease Development

- **Disease Triangle:** Disease development depends on the interaction of three components - a susceptible host plant, a virulent pathogen, and conducive environmental conditions.
- **Infection Process:** Pathogens enter the plant through wounds, natural openings, or direct penetration, and they can then multiply and spread within the plant.
- **Symptoms and Diagnosis:** Plant pathologists identify diseases by studying the symptoms and signs, conducting laboratory tests, and using molecular techniques. Symptoms can vary widely, including wilting, yellowing, necrosis, and deformities.

Disease Management

- **Cultural Practices:** Techniques like crop rotation, sanitation, and proper irrigation can help reduce the spread of diseases.
- **Chemical Control:** The use of fungicides, bactericides, and other chemicals can manage some plant diseases.
- **Biological Control:** Beneficial microorganisms or predators can be used to control pathogens.
- **Host Resistance:** Developing resistant plant varieties through breeding and genetic engineering is an effective long-term strategy.

Plant pathology plays a vital role in ensuring the health and productivity of plant populations, and it continues to evolve with advancements in technology and research. It's a dynamic field with important implications for food security and environmental conservation.

Objectives of Plant Pathology

Plant pathology is the scientific study of plant diseases, and its objectives are multifaceted, encompassing both understanding and managing plant diseases. The main objectives of plant pathology include:

Disease Identification and Diagnosis

Identify and classify plant diseases, including their causes (pathogens, abiotic factors, *etc.*).

Develop diagnostic methods and tools to accurately identify diseases.

Understanding Disease Causation

Investigate the biology and genetics of plant pathogens (*e.g.*, fungi, bacteria, viruses, nematodes).

Study the interactions between pathogens and plants at the molecular and cellular levels.

Epidemiology

Study the factors that influence the spread and severity of plant diseases, including environmental conditions and host-pathogen interactions.

Predict disease outbreaks and establish control strategies based on epidemiological data.

Disease Management

Develop and implement strategies for the prevention, control, and management of plant diseases.

Explore cultural, biological, chemical, and genetic methods for disease management.

Breeding for Disease Resistance

Develop disease-resistant plant varieties through selective breeding and genetic engineering.

Enhance crop yield and quality by reducing the impact of diseases.

Environmental Stewardship

Promote sustainable and environmentally friendly practices for disease management, reducing the need for chemical pesticides.

Food Security

Contribute to global food security by minimizing crop losses due to plant diseases.

Develop strategies to protect food crops from diseases that can affect human and animal health.

Research and Education

Advance the scientific understanding of plant diseases and their management through research.

Educate and train plant pathologists, farmers, and the public about plant diseases and their control.

Plant Health Regulation

Develop and enforce regulations and policies related to plant health and quarantine measures to prevent the introduction and spread of diseases.

Disease Forecasting

Develop early warning systems for disease outbreaks to enable timely control measures.

Utilize weather and environmental data for predictive modeling of disease occurrence.

International Collaboration

Collaborate with other countries and international organizations to address global plant health challenges and exchange information on disease management.

Plant pathology plays a crucial role in agriculture, horticulture, forestry, and natural ecosystems, as it helps protect plants from diseases and ensures the sustainable production of food, fiber, and ornamental crops. By achieving these objectives, plant pathologists contribute to global food security, environmental sustainability, and the overall well-being of society.

Group of Plant Pathogens

Plant pathogens, collectively referred to as phytopathogens, encompass a diverse array of organisms and agents that pose formidable challenges to the health and vitality of plants. These adversaries include fungi, bacteria, viruses, nematodes, oomycetes, and other microorganisms, each with its own distinctive biology and mode of attack. Their insidious presence in the botanical realm can lead to a myriad of diseases, ranging from the telltale signs of wilting leaves, unsightly blotches, and discoloration to more subtle, underlying stresses that weaken the plant's ability to thrive. Beyond these biological antagonists, environmental factors, such as abiotic stresses and nutrient imbalances, can further exacerbate plant susceptibility to disease. Understanding the vast and intricate world of plant pathogens is not only pivotal for safeguarding agricultural and horticultural yields but is also essential for protecting the integrity of ecosystems, ensuring global food security, and fostering sustainable practices in an ever-evolving world. In this rich tapestry of botanical health and ailment, the study of plant pathology plays a central role in unraveling the secrets of these pathogens, their interactions with their plant hosts, and the development of strategies to mitigate their impact.

Plant pathogens can be grouped into several categories based on their biological characteristics and the types of diseases they cause. The main groups of plant pathogens include:

Fungi

Fungi are eukaryotic organisms that include microorganisms such as yeasts, moulds and mushrooms. These organisms are classified under kingdom fungi. The organisms found in Kingdom fungi contain a cell wall and are omnipresent. They are classified as heterotrophs among the living organisms.

- **Eukaryotic Organisms:** Fungi are eukaryotes, which means they have complex cells with a well-defined nucleus and various organelles, just like plants, animals, and protists.
- **Diversity:** The kingdom Fungi is incredibly diverse, encompassing a wide range of organisms. Some common examples include yeasts, moulds, and mushrooms, but there are thousands of different species within this kingdom, each with its own unique characteristics.
- **Cell Wall:** Fungi are known for having cell walls. These cell walls are primarily composed of a complex sugar called chitin, which is distinct from the cellulose found in plant cell walls.
- **Heterotrophs:** Fungi are heterotrophic, meaning they obtain their nutrients by absorbing organic material from their surroundings. They are unable to photosynthesize like plants do. Instead, they secrete enzymes that break down organic matter, such as dead plants or animals, and then absorb the resulting nutrients.
- **Omnipresent:** Fungi are found almost everywhere on Earth. They can thrive in various environments, from the depths of the ocean to the high mountains, and can even colonize extreme environments like deserts and polar regions.

Fungi play essential roles in ecosystems by decomposing organic matter, forming symbiotic relationships with plants (mycorrhizae), and serving as food sources for various organisms. They are also used by humans in various ways, such as in the production of food (yeast for baking and brewing), antibiotics (like penicillin), and as a source of various industrial enzymes.

Bacteria

Bacterial pathogens can infect various plant parts, leading to diseases such as fire blight, bacterial wilt, and citrus canker. Bacteria are a diverse group of single-celled microorganisms that belong to the domain Bacteria. They are one of the three domains of life, with the other two being Archaea and Eukarya

- **Prokaryotic Cells:** Bacteria are prokaryotes, which means they lack a true nucleus and membrane-bound organelles. Their genetic material is contained in a single, circular DNA molecule located in the nucleoid region of the cell.
- **Shapes and Sizes:** Bacteria come in various shapes, including spheres (cocci), rods (bacilli), and spirals. They also vary in size, with some being as small as 0.5 micrometers and others reaching a few micrometers in length.
- **Cell Walls:** Most bacteria have cell walls, which provide structural support and help protect the cell. The composition of the cell wall can vary; for example, Gram-positive bacteria have a thick layer of peptidoglycan, while Gram-negative bacteria have a thinner peptidoglycan layer surrounded by an outer membrane.
- **Nutrition:** Bacteria can be classified based on their nutritional requirements. Some are autotrophic, meaning they can produce their own food through processes like photosynthesis or chemosynthesis, while others are heterotrophic, relying on organic compounds from their environment.
- **Reproduction:** Bacteria reproduce asexually through binary fission, where one bacterial cell divides into two identical daughter cells. This process allows for rapid population growth.
- **Ubiquitous:** Bacteria are incredibly widespread and can be found in almost every environment on Earth, from the depths of the ocean to the upper atmosphere. They play crucial roles in biogeochemical cycles, such as the carbon and nitrogen cycles.
- **Beneficial and Harmful:** Bacteria have both beneficial and harmful roles. Beneficial bacteria are involved in processes like nitrogen fixation (converting atmospheric nitrogen into a usable form for plants), and they are used in food production (*e.g.*, yogurt and cheese) and in biotechnology. Harmful bacteria can cause diseases in humans, animals, and plants and are responsible for various infections.
- **Antibiotics:** The discovery of antibiotics, such as penicillin, revolutionized medicine by providing effective treatments for bacterial infections. However, overuse and misuse of antibiotics have led to the development of antibiotic-resistant strains of bacteria, posing a significant public health concern.

Understanding bacteria is essential in fields such as microbiology, medicine, and biotechnology, as they are involved in various aspects of human life and have a significant impact on ecosystems and the environment.

Viruses

Plant viruses are submicroscopic infectious agents that can cause a wide range of symptoms, including mosaic patterns, yellowing, and stunted growth. Examples include Tobacco mosaic virus and Potato virus Y. Viruses are unique and fascinating infectious agents that are distinct from both bacteria and other microorganisms. They are not classified within any of the three domains of life (Bacteria, Archaea, and Eukarya) because they lack the characteristics of living organisms.

- ☆ **Non-Living Entities:** Viruses are considered non-living because they lack essential cellular structures and cannot carry out metabolic processes on their own. They do not have cells, organelles, or the ability to reproduce without a host cell.
- ☆ **Genetic Material:** Viruses consist of genetic material, either DNA or RNA, surrounded by a protein coat called a capsid. Some viruses also have an outer lipid envelope derived from the host cell's membrane.
- ☆ **Obligate Intracellular Parasites:** Viruses are obligate intracellular parasites, which means they cannot replicate or carry out their life cycle outside of a host cell. They must infect a living host cell to reproduce.
- ☆ **Host Specificity:** Each virus has a specific host range, which may include particular species, cell types, or even individual organisms. This specificity is due to the interaction between viral surface proteins and host cell receptors.
- ☆ **Replication:** When a virus infects a host cell, it hijacks the host's cellular machinery to replicate its genetic material and produce new virus particles. This process can lead to the destruction of the host cell and the release of newly formed viruses.
- ☆ **Diverse Shapes and Sizes:** Viruses come in a wide range of shapes and sizes. Some are simple and geometric in shape, while others are more complex. They can be as small as 20 nanometers in diameter or much larger.
- ☆ **Human Health Impact:** Viruses are responsible for many diseases in humans, such as the common cold, influenza, HIV/AIDS, COVID-19, and more. Vaccines have been developed to prevent some viral infections, while antiviral drugs are used to treat others.
- ☆ **Biotechnological Applications:** Some viruses have been harnessed for biotechnological applications, including gene therapy and the development of viral vectors for delivering therapeutic genes into host cells.
- ☆ **Bacteriophages:** Bacteriophages, or phages, are viruses that infect and replicate within bacterial cells. They have been studied for their potential in controlling bacterial infections and in biotechnology.

- ☆ **Evolution and Origins:** The origin of viruses is still a subject of scientific debate. Some theories suggest that viruses may have evolved from genetic elements or from early cellular life forms.

Viruses are a significant area of study in virology, and understanding their biology is crucial for both human health and biotechnology. They represent a unique and complex group of infectious agents that have a profound impact on life on Earth.

Nematodes

Plant-parasitic nematodes are microscopic roundworms that infect plant roots, causing diseases like root-knot nematode and cyst nematode infection.

Nematodes, often referred to as roundworms, are a diverse group of multicellular, unsegmented worms that belong to the phylum Nematoda. They are found in a wide range of environments, from soil and freshwater to marine ecosystems.

- ☆ **Body Structure:** Nematodes have long, slender, unsegmented bodies that are typically cylindrical in shape. They range in size from microscopic to several centimeters long. Their bodies are covered by a tough, flexible cuticle that is periodically molted as they grow.
- ☆ **Habitats:** Nematodes are incredibly abundant and can be found in various habitats, including soil, sediments, water, and even as parasites in plants and animals. They play essential roles in ecosystem nutrient cycling, particularly in soil ecosystems.
- ☆ **Diversity:** Nematodes are an incredibly diverse group of organisms, with over 25,000 described species. However, it is estimated that there may be hundreds of thousands of nematode species yet to be discovered.
- ☆ **Feeding:** Nematodes exhibit a range of feeding strategies. Some are herbivores, feeding on plant roots or other organic material. Others are carnivorous, preying on smaller organisms like bacteria, fungi, and even other nematodes. Some nematodes are parasites, causing diseases in plants, animals, and humans.
- ☆ **Reproduction:** Nematodes reproduce sexually, with separate male and female individuals. Fertilized eggs are laid and then hatch into juvenile nematodes. Nematodes grow through a series of molts as they mature into adults.
- ☆ **Ecological Roles:** Nematodes are essential in nutrient cycling and decomposition in ecosystems. They help break down organic matter, releasing nutrients that can be taken up by plants and other organisms. Some nematodes also act as predators, controlling the populations of other microscopic organisms.

- **Economic Importance:** While many nematodes play beneficial roles in ecosystems, some are economically significant pests. Plant-parasitic nematodes can damage agricultural crops, causing reduced yields and economic losses.
- **Research and Study:** Nematodes are used as model organisms in scientific research. *Caenorhabditis elegans*, a small, transparent nematode, is one of the most extensively studied organisms in biology due to its simple nervous system and rapid life cycle. It has provided valuable insights into genetics, development, and neuroscience.
- **Medical Relevance:** Some nematodes can be parasitic to humans and animals, causing diseases like filariasis and hookworm infections. These parasites can have significant public health impacts in certain regions.

Understanding the diversity and biology of nematodes is important not only for ecological and agricultural purposes but also for research in various scientific disciplines, including biology, genetics, and parasitology.

Oomycetes

Oomycetes are water molds that can cause diseases such as late blight in potatoes and downy mildew in various crops. Oomycetes, often referred to as water molds, are a group of microorganisms that were once classified as fungi due to their filamentous growth and similar ecological niches. However, they are now recognized as a distinct group within the kingdom Stramenopila. Oomycetes are primarily known for their devastating impact on plants, especially in agriculture, as many species are responsible for various plant diseases.

- **Cell Structure:** Oomycetes have cell walls made of cellulose, which is a key distinction from true fungi that have chitin in their cell walls.
- **Habitat:** They are predominantly aquatic or thrive in damp environments. Oomycetes can be found in soil, freshwater habitats, and even in marine environments.
- **Reproduction:** Oomycetes reproduce both sexually and asexually. Asexual reproduction typically involves the formation of sporangia, which release motile zoospores. Sexual reproduction may involve the fusion of specialized sexual structures.
- **Pathogens:** Many oomycetes are notorious plant pathogens. One of the most well-known examples is *Phytophthora infestans*, the causal agent of the Irish Potato Famine in the 19th century. This water mold was responsible for the widespread destruction of potato crops.
- **Human and Animal Pathogens:** While they are primarily plant pathogens, some oomycetes can infect fish, amphibians, and other

animals, including humans. For example, the genus Pythium includes species that can cause infections in humans.

- ☆ **Decomposers:** Some oomycetes play a role in breaking down dead organic matter, contributing to nutrient recycling in ecosystems.
- ☆ **Saprobes:** Besides being pathogens, some oomycetes are saprobic, meaning they feed on dead or decaying organic material.
- ☆ **Phylogenetic Placement:** Molecular studies have shown that oomycetes are more closely related to brown algae, diatoms, and other stramenopiles than they are to true fungi.
- ☆ **Ecological Impact:** Oomycetes can have a significant ecological impact, both as pathogens and as regulators of nutrient cycling in ecosystems.
- ☆ **Control:** Managing oomycete diseases in agriculture often involves the use of fungicides or other control measures. Understanding the biology and life cycle of these pathogens is essential for effective disease management.

Oomycetes are a group of microorganisms with some similarities to fungi but distinct characteristics, such as cellulose cell walls. They are known for their plant pathogenicity and can cause devastating diseases in agriculture. Understanding oomycetes is important for managing these diseases and their ecological impact.

Phytoplasmas and Spirplasmas

These are cell-wall-less bacteria that infect plants and are transmitted by insects. They cause diseases like aster yellows and citrus stubborn disease.

Phytoplasmas and spiroplasmas are both groups of microorganisms that belong to the class Mollicutes, a class of bacteria that lack a cell wall. They are known for their unique features and association with plant and insect diseases.

Phytoplasmas

- ☆ **Classification:** Phytoplasmas are a type of mollicute that are obligate plant pathogens. They were formerly classified as mycoplasma-like organisms.
- ☆ **Hosts:** Phytoplasmas primarily infect plants and are responsible for a wide range of plant diseases. These diseases include "yellows," which are characterized by yellowing and other symptoms in plant tissues.
- ☆ **Transmission:** Phytoplasmas are typically transmitted by insect vectors, such as leafhoppers and planthoppers. The insects feed on infected plants and then transmit the pathogen to healthy plants during subsequent feeding.
- ☆ **Symptoms:** Phytoplasma infections can cause a variety of symptoms in plants, including leaf yellowing, stunting, witches' broom (abnormal proliferation of branches), and other deformities.

- **Economic Impact:** Phytoplasma diseases can have a significant economic impact in agriculture as they can affect a wide range of crop plants.
- **Control:** Managing phytoplasma diseases often involves controlling the insect vectors and removing infected plants to prevent the spread of the pathogen.

Spiroplasmas

- **Classification:** Spiroplasmas are also a group of mollicutes, but they are distinct from phytoplasmas and have a helical or spiral shape.
- **Hosts:** Spiroplasmas infect a variety of hosts, including plants, insects, and animals. In plants, they can cause diseases, while in insects, they are often found as commensals (beneficial bacteria) or parasites.
- **Transmission:** Spiroplasmas can be transmitted by insect vectors, much like phytoplasmas. They are often found in the phloem (the plant's vascular tissue), and insect vectors that feed on the phloem can acquire and transmit these bacteria.
- **Animal Diseases:** Some spiroplasmas are associated with diseases in animals, particularly livestock and poultry. They can cause conditions like "corn stunt" in cattle and other ailments in birds.
- **Research:** Spiroplasmas have been studied in the context of symbiosis and parasitism, making them of interest in understanding these interactions in various host organisms.

Both phytoplasmas and spiroplasmas are of significance due to their roles in various diseases and interactions with plants, insects, and animals. Understanding their biology and the mechanisms of transmission is essential for disease management and control measures, particularly in agriculture and vector-borne diseases.

Mycoplasma-Like Organisms (MLOs)

MLOs are also cell-wall-less bacteria that can cause diseases like peach X-disease and tomato big bud.

Mycoplasma-Like Organisms (MLOs) are a group of microorganisms that were initially thought to be a type of mycoplasma, a class of bacteria that lack a cell wall and are among the smallest free-living organisms. However, it's now understood that MLOs are not true mycoplasmas but rather a diverse group of cell-wall-less bacteria belonging to various genera.

MLOs are notable for their small size and unique characteristics, and they have been associated with a variety of diseases in plants and insects. Some MLOs are responsible for plant diseases, particularly in crops like grapes, where they cause a condition known as "flavescence dorée" or "golden yellow disease," and in citrus trees, where they are responsible for "huanglongbing" or citrus greening disease. These diseases can be devastating to agriculture, leading to significant economic losses.

In addition to plant diseases, MLOs have been found in insects, particularly leafhoppers and psyllids, which serve as vectors for MLO transmission. When an infected insect feeds on a plant, it can transmit MLOs, leading to the spread of diseases.

Researchers are still studying MLOs and the diseases they cause to better understand their biology and develop strategies for controlling them. The use of insecticides and the development of disease-resistant plant varieties are among the approaches used to manage MLO-associated diseases in agriculture.

It's important to note that the field of microbiology and our understanding of various microorganisms is continually evolving.

Virophages

Virophages are small viruses that infect larger plant viruses, affecting the spread and impact of certain plant diseases. Virophages are small viruses that are parasitic on other viruses, particularly giant viruses, and they are characterized by their ability to replicate within the host cells of these giant viruses. The discovery of virophages has added an intriguing layer to the complex world of viruses, highlighting the intricate relationships that exist between different types of viruses.

- ☆ **Discovery:** Virophages were first discovered in 2008 in a study of a giant virus known as Acanthamoeba polyphaga mimivirus (APMV). Researchers found smaller viruses inside APMV-infected amoeba cells, and these smaller viruses were termed "virophages."
- ☆ **Relationship with Giant Viruses:** Virophages are not parasites of cellular organisms like animals or plants; instead, they parasitize giant viruses. Giant viruses are a unique group of viruses that have large genomes and complex structures. Virophages co-infect the same host cell as the giant virus and utilize the host's machinery to replicate.
- ☆ **Replication and Interference:** Virophages are essentially viral parasites of giant viruses. They replicate within the same host cell, potentially interfering with the replication of the giant virus. This interference can reduce the productivity of the giant virus and benefit the host cell.
- ☆ **Biological Significance:** Virophages have raised questions about the co-evolution and interactions between different types of viruses. They may influence the dynamics of viral communities within environmental and biological systems. Understanding virophages and their roles in ecosystems is still an active area of research.
- ☆ **Examples:** The first discovered virophage, Sputnik, was found in association with APMV. Subsequently, other virophages have been

identified, such as Mavirus and Zamilon, which are associated with different giant viruses.

- ☆ **Implications:** Virophages have raised interest in the potential use of these viruses for various applications, such as biocontrol agents in combating giant virus infections or as tools in molecular biology and biotechnology.

The discovery of virophages has expanded our understanding of the complexity of viral interactions and their potential impacts on ecosystems and biological systems. It underscores the intricate relationships and competition that exist within the viral world, with smaller viruses affecting the replication and success of larger ones.

Algae and Protozoa

Some green algae and protozoa can be plant pathogens, causing diseases like algal leaf spot and slime molds. Algae and protozoa are both diverse groups of microorganisms, but they belong to different biological kingdoms and have distinct characteristics.

Algae

- ☆ **Classification:** Algae are a diverse group of photosynthetic microorganisms that belong to the kingdom Protista or Plantae, depending on the classification system. They are primarily classified based on their pigments and cell wall composition.
- ☆ **Photosynthesis:** Algae are autotrophic organisms, which means they can produce their own food through photosynthesis. They contain chlorophyll and other pigments that enable them to capture sunlight and convert it into energy.
- ☆ **Cell Structure:** Algal cells can vary in size and complexity. Some algae are unicellular, while others are multicellular. They have a cell wall made of cellulose, similar to plant cells.
- ☆ **Habitat:** Algae are found in various aquatic environments, including freshwater and marine ecosystems. They can also grow in moist terrestrial habitats.
- ☆ **Importance:** Algae are important for aquatic ecosystems as they are a primary food source for many aquatic organisms. They are also used in various industrial applications, including the production of food, biofuels, and pharmaceuticals.

Protozoa

- ☆ **Classification:** Protozoa are a diverse group of single-celled microorganisms that belong to the kingdom Protista. They are classified based on their method of locomotion and other characteristics.

- **Nutrition:** Protozoa can be autotrophic or heterotrophic. Autotrophic protozoa can photosynthesize, but the majority of them are heterotrophic, meaning they obtain their nutrients by ingesting other microorganisms or organic matter.
- **Cell Structure:** Protozoa are typically unicellular and lack a cell wall. Their cell membrane surrounds the cell and controls the exchange of substances with the environment.
- **Locomotion:** Protozoa are known for their various methods of locomotion. Some move using cilia, flagella, or pseudopodia (temporary extensions of the cell membrane), while others are non-motile.
- **Habitat:** Protozoa can be found in a wide range of habitats, including aquatic environments, soil, and the gastrointestinal tracts of animals. They are also commonly found in wastewater treatment processes.
- **Importance:** Protozoa play important roles in nutrient cycling and food chains, both in aquatic and terrestrial ecosystems. Some protozoa are also responsible for causing diseases in humans and other animals.

In summary, while both algae and protozoa are microorganisms, they differ in terms of their classification, nutritional modes, cell structure, and ecological roles. Algae are primarily photosynthetic, while protozoa can be either autotrophic or heterotrophic, with diverse methods of locomotion.

Parasitic Plants

Some plants, such as dodder and broomrape, are parasitic and directly attach to host plants to extract nutrients, causing damage. Parasitic plants are a unique group of plants that obtain some or all of their nutrients from other living plants. Unlike typical plants, which use photosynthesis to generate their own food from sunlight, water, and carbon dioxide, parasitic plants have evolved specialized adaptations to tap into the resources of their host plants.

Types of Parasitic Plants

- **Holoparasites:** These plants do not have chlorophyll and are entirely dependent on their host for nutrients. They usually attach to the host plant's roots and can be internal or external parasites.
- **Hemiparasites:** Hemiparasitic plants have some chlorophyll and are capable of photosynthesis, but they also obtain nutrients from their host plants through specialized structures, such as haustoria (modified roots that penetrate the host plant).

Haustoria: Many parasitic plants have specialized structures called haustoria, which are used to penetrate the host plant's vascular system, allowing the parasite to extract water, nutrients, and even sugars from the host.

Host Specificity: Some parasitic plants are highly host-specific, meaning they can only parasitize specific plant species, while others are more generalist and can parasitize a wide range of host plants.

Examples of Parasitic Plants

- **Dodder (*Cuscuta* spp.):** Dodder is a leafless, twining parasitic plant that wraps itself around the host plant and forms haustoria to extract nutrients.
- **Mistletoe (*Viscum* album and others):** Mistletoe is a hemiparasitic plant that grows on trees and shrubs. It has green leaves and conducts some photosynthesis but also relies on its host for water and minerals.
- **Broomrape (*Orobanche* spp.):** Broomrape is a holoparasitic plant that lacks chlorophyll and attaches to the roots of various host plants to obtain all its nutrients.
- **Indian Paintbrush (*Castilleja* spp.):** Indian paintbrush is a semi-parasitic plant that attaches to the roots of nearby plants and obtains water and nutrients.

Ecological Role: Parasitic plants play a role in their ecosystems by regulating the populations of their host plants and influencing plant community dynamics. They can also affect nutrient cycling in the environment.

Control and Management: Some parasitic plants can be detrimental to crops and natural ecosystems. Various strategies, such as herbicides, biological control, and cultural practices, may be used to manage parasitic plants in agriculture and conservation efforts.

Parasitic plants have evolved unique mechanisms to exploit other plants for their survival. They are a fascinating and diverse group of plants with a range of ecological interactions and adaptations that continue to be of interest to botanists and ecologists.

Prions

Prions are infectious proteins that can cause diseases in some plants, affecting their growth and development. Prions are unique, infectious agents composed of misfolded proteins. They are associated with several rare and fatal neurodegenerative diseases in humans and other animals. The name "prion" is derived from "proteinaceous infectious particle."

Key characteristics and properties of prions include:

- **Misfolded Proteins:** Prions are typically a misfolded form of a normal cellular protein called PrP (prion protein). When the normal PrP protein misfolds into an abnormal shape, it can serve as a template to induce other normal PrP proteins to misfold in the same way.

- **Infectious:** Prions are infectious in the sense that they can induce the misfolding of other proteins. This is a self-perpetuating process that leads to the accumulation of misfolded proteins in the brain.
- **Neurodegenerative Diseases:** Prion diseases, or transmissible spongiform encephalopathies (TSEs), are characterized by the accumulation of misfolded PrP proteins in the brain, leading to progressive neurological dysfunction. Examples of prion diseases in humans include Creutzfeldt-Jakob disease (CJD), variant Creutzfeldt-Jakob disease (vCJD), and kuru. In animals, examples include bovine spongiform encephalopathy (BSE or "mad cow disease") and scrapie in sheep.
- **Lack of Genetic Material:** Unlike viruses or bacteria, prions do not contain genetic material like DNA or RNA. Their infectious nature is solely based on the ability to induce conformational changes in other proteins.
- **Resistance to Heat and Disinfection:** Prions are highly resistant to standard sterilization methods, including heat and chemical disinfectants, which makes them particularly challenging to eliminate.
- **No Known Cure:** Currently, there is no cure for prion diseases. These conditions are often rapidly progressive and fatal.

The exact mechanisms by which prions induce normal proteins to misfold and cause neurodegeneration are not fully understood. The study of prions remains an active area of research, and understanding their biology is crucial for developing potential treatments and interventions for prion diseases. Additionally, because of their infectious nature and resistance to traditional sterilization methods, prions have raised concerns in the context of food safety and medical procedures.

Each group of plant pathogens may have different characteristics, modes of transmission, and effects on plants, necessitating specific strategies for their control and management. Plant pathologists study these pathogens to understand their biology, ecology, and genetics, which helps in developing effective disease management strategies.

Economic Importance of Plant Diseases

Plant diseases have significant economic implications that affect both agricultural and horticultural sectors, as well as the broader economy. Here are some of the key economic impacts of plant diseases:

- **Crop Losses:** Plant diseases can cause substantial losses in crop yields. Crop diseases can lead to reduced harvests, lower crop quality, and increased production costs due to the need for disease management practices, such as the application of pesticides and fungicides.

- **Reduced Crop Value:** Diseased crops often have lower market value due to lower quality and visual defects. This can lead to decreased profits for farmers and reduced revenue for the agriculture industry.
- **Increased Production Costs:** Controlling and managing plant diseases can be expensive. Farmers often have to invest in pesticides, fungicides, and other treatments to protect their crops. These costs can increase the overall cost of production, which can, in turn, raise food prices for consumers.
- **Trade Restrictions:** Plant diseases can result in trade restrictions. Countries may impose import bans or strict quarantine measures to prevent the introduction and spread of plant diseases, disrupting international trade and causing economic losses.
- **Impact on Food Security:** Plant diseases can threaten food security by reducing the availability of crops. This can lead to food shortages, rising food prices, and, in some cases, food crises, especially in regions highly dependent on agriculture.
- **Loss of Income for Farmers:** When plant diseases devastate crops, farmers' incomes are directly impacted. This can lead to financial hardships, reduced investments in the agricultural sector, and decreased economic activity in rural areas.
- **Job Losses:** The agricultural and food processing industries are major employers in many regions. Plant diseases can result in job losses as reduced production leads to less demand for labor in these sectors.
- **Supply Chain Disruptions:** Plant diseases can disrupt supply chains, affecting not only farmers but also agribusinesses, transportation, and food retailers. This can result in economic losses throughout the entire supply chain.
- **Research and Development Costs:** Governments and private industries invest substantial resources in research and development to combat plant diseases. The costs associated with developing disease-resistant crop varieties and effective control measures can be substantial.
- **Environmental Impact:** Some disease management practices, such as the use of pesticides, can have negative environmental consequences. The costs of environmental damage and remediation can also be significant.
- **Loss of Biodiversity:** Plant diseases can impact natural ecosystems, leading to the loss of plant species and biodiversity, which can have economic consequences for sectors like ecotourism and pharmaceuticals.
- **Risk Management Costs:** Businesses and governments need to allocate resources to risk management and disaster preparedness to address the potential economic impacts of plant diseases.

Plant diseases have a far-reaching economic impact, affecting agricultural productivity, food security, trade, employment, and the overall well-being of economies. Preventative measures, early detection, and disease management strategies are essential for mitigating these economic consequences. Additionally, ongoing research and investments in disease-resistant crop varieties and sustainable farming practices are crucial for reducing the economic burden of plant diseases.

Chapter 2

History of Plant Pathology

History: An Introduction

The history of plant pathology is a captivating journey that chronicles humanity's evolving understanding of the diseases that afflict plants, and the relentless quest to combat these agricultural scourges. In the annals of plant pathology, early civilizations grappled with mysterious plant maladies, often attributing them to supernatural forces or environmental factors. However, it was during the 17th to 19th centuries that the foundations of modern plant pathology were laid, as scientists and naturalists began systematically studying plant diseases. Notable advancements came in the 19th century with Agostino Bassi's discovery of a microscopic pathogen causing silkworm disease. This marked one of the initial instances where an infectious microorganism was linked to a plant or animal ailment.

The mid-19th century brought a watershed moment in plant pathology with the devastating Irish Potato Famine, triggered by the potato late blight caused by *Phytophthora infestans*. This humanitarian disaster underscored the urgent need for scientific investigation into plant diseases. Furthermore, the late 19th century witnessed the development of the germ theory of disease by Louis Pasteur and Robert Koch, which greatly advanced the understanding of infectious agents in both human and plant diseases.

As the 20th century dawned, plant pathology gained prominence, with bacteriologists and virologists identifying bacteria and viruses as culprits of plant ailments. This era also witnessed the development of diagnostic techniques and chemicals for controlling diseases. The mid-20th century, known as the Green Revolution, saw an intensive focus on the development of disease-resistant crop varieties and the use of fungicides and pesticides to protect crops, ultimately increasing food production.

The late 20th century brought ground breaking advancements in biotechnology and genetic engineering, enabling the creation of genetically modified (GM) crops with resistance to pests and diseases. The 21st century has seen a renewed emphasis on sustainable agriculture, with an increased focus on environmentally friendly methods, biological control, and a reduced reliance on chemical pesticides. As the world grapples with the challenges posed by climate change and global trade, new plant pathogens, pests, and diseases continually emerge, necessitating ongoing research and innovative strategies to safeguard global food security.

The history of plant pathology represents the transition from ancient folklore to a highly specialized scientific discipline. Plant pathologists now play a critical role in deciphering, preventing, and managing diseases that imperil crops, ensuring food production and agricultural sustainability on a global scale. This rich history reminds us of the importance of continuous innovation and collaboration to address the ever-evolving challenges in safeguarding our crops and food supply.

Brief Overview of the Key Developments in the History of Plant Pathology

The history of plant pathology is a fascinating journey that involves the study of plant diseases, their causes, and methods for their prevention and control. Here's a brief overview of the key developments in the history of plant pathology:

Early Observations (Pre-17th Century)

The pre-17th century period was marked by early observations and understandings of plant diseases, representing a foundational stage in the history of plant pathology. In this era, explanations for crop ailments were often rooted in supernatural or mystical beliefs. People attributed plant diseases to factors such as bad omens, evil spirits, or celestial events. These superstitious interpretations were further ingrained in various cultures and societies, where rituals and offerings were conducted to appease these perceived forces and protect crops.

However, amid these superstitions, some individuals began to make more systematic and empirical observations of plant diseases. For instance, Theophrastus, often regarded as the "Father of Botany," made notable contributions in ancient Greece, documenting the symptoms of various plant ailments. His observations of plant diseases, although somewhat rudimentary by today's standards, represented an early attempt to categorize and describe these conditions based on visible symptoms.

In ancient China, records of plant diseases date back to approximately 3000 BCE, with descriptions of outbreaks and crop losses attributed to specific diseases. Chinese agriculture also emphasized the importance of crop rotation and the use of green manure to enhance soil fertility and reduce disease pressure.

In other parts of the world, ancient civilizations such as the Romans, Egyptians, and Indians documented their observations of plant diseases and attempted rudimentary control measures. These early efforts laid the groundwork for later developments in plant pathology, as they marked the first attempts to understand and manage the diseases that affected agriculture.

While these early observations were often steeped in mysticism and limited by a lack of scientific understanding, they provided the foundation for future generations to systematically study plant diseases and ultimately gave rise to the field of plant pathology as a distinct and scientific discipline.

Plant diseases have plagued agricultural societies throughout history. However, systematic study and documentation of plant diseases began relatively late. Early farmers and gardeners had some knowledge of common plant ailments, but the understanding of their causes was limited. During the pre-17th century period, early observations of plant diseases were largely rooted in anecdotal and experiential knowledge. Agriculture and horticulture practices had been established for millennia, but the understanding of the causes of plant diseases remained limited, and explanations often veered into the realm of folklore and superstition.

Several Key Developments Characterized Early Observations (Pre-17th Century)

- **Anecdotal Knowledge:** Farmers, gardeners, and herbalists possessed a wealth of empirical knowledge about plant diseases. They recognized common symptoms such as wilting, discoloration, and deformities, but the causes were often attributed to vague factors like "ill humors" or unfavorable cosmic influences.
- **Superstition and Mysticism:** In the absence of scientific understanding, plant diseases were sometimes attributed to supernatural forces or curses. For example, the concept of "plant witches" or malevolent spirits that harmed crops was prevalent in various cultures.
- **Early Attempts at Disease Management:** While the true nature of pathogens was not understood, some practical methods for managing plant diseases emerged. Crop rotation and fallowing were used to mitigate soilborne diseases, and protective practices like burning infected plants were employed.
- **Limited Documentation:** The limited documentation and absence of systematic records from this era make it challenging to reconstruct a comprehensive history of pre-17th century plant pathology. Instead,

knowledge was largely passed down orally and through practical experience.

The pre-17th century period in plant pathology was characterized by a lack of scientific understanding and the reliance on tradition, folklore, and anecdotal knowledge to explain and manage plant diseases. It marked the earliest, rudimentary steps in humanity's engagement with the complex world of plant pathogens, eventually leading to more systematic and scientific approaches in subsequent centuries.

The Birth of Modern Plant Pathology (17th to 19th Century)

The foundation of modern plant pathology was laid during the 17th to 19th centuries. Notable developments during this period include the work of Italian physician Agostino Bassi, who in the early 19th century, demonstrated that a silkworm disease was caused by a microscopic organism. This was one of the first documented instances of an infectious microorganism causing a plant or animal disease.

The birth of modern plant pathology during the 17th to 19th centuries represents a transformative period in the understanding of plant diseases. This era witnessed a gradual shift from supernatural and mystical explanations for crop ailments to more systematic and scientific approaches. Pioneering work by individuals like Italian polymath Agostino Scopoli and the Swedish botanist Carl Linnaeus laid the groundwork for the classification and identification of plant diseases. The emergence of microscopy in the 17th century allowed for the observation of pathogenic agents such as fungi and bacteria, paving the way for the study of their roles in plant diseases. Additionally, the devastating impact of major plant epidemics, such as the late blight in potatoes in the mid-19th century, served as catalysts for scientific inquiry into the nature and causes of plant diseases.

This period also saw the development of key concepts, including the distinction between biotic (living) and abiotic (non-living) causes of diseases. Prominent plant pathologists like Anton de Bary, known as the "father of modern mycology," furthered the understanding of the role of fungi in plant diseases. The birth of modern plant pathology represented a fundamental shift towards the application of scientific methodology, including the use of the scientific method and controlled experiments to investigate the causes, transmission, and management of plant diseases. These developments laid the foundation for the field of plant pathology as a distinct scientific discipline, enabling the development of strategies to combat and prevent plant diseases and contributing to the advancement of global agriculture.

The period from the 17th to the 19th century marked the birth of modern plant pathology, as scientists and naturalists began to systematically study plant diseases, laying the foundation for a more scientific understanding of these afflictions.

Several Key Developments Characterized this Era

- **Emergence of Systematic Observation:** During the 17th century, there was a shift towards more systematic and scientific observation of plant diseases. Naturalists and botanists began to document the symptoms of various plant ailments and their impact on agriculture and horticulture.
- **Publication of Early Works:** In the 18th century, notable publications by scientists such as Anton de Bary and Pierre-Augustin Broussonet contributed to the emerging field of plant pathology. De Bary, in particular, is often regarded as a pioneer in the study of plant diseases.
- **Discovery of Fungal Pathogens:** The 19th century saw significant progress in identifying the role of fungi as causal agents of plant diseases. Notable discoveries included the work of Miles Joseph Berkeley, who described various plant pathogenic fungi.
- **Understanding Infectious Agents:** Researchers began to understand that microscopic organisms were responsible for causing plant diseases. This was a fundamental shift in thinking and laid the groundwork for modern plant pathology.
- **Potato Late Blight:** The devastating Irish Potato Famine in the mid-19th century, caused by the oomycete pathogen *Phytophthora infestans*, brought international attention to the study of plant diseases and underscored the need for scientific investigation and disease management.
- **Microscopy and Improved Research Tools:** Advances in microscopy and research tools allowed scientists to better examine plant tissues and the microorganisms responsible for diseases. This led to more accurate identification of pathogens.
- **Bacterial Diseases:** Towards the end of the 19th century, researchers began to identify bacterial pathogens responsible for plant diseases. This was a critical step in understanding the diversity of plant pathogens.

The period from the 17th to the 19th century represented a pivotal time in the history of plant pathology. It marked the transition from anecdotal knowledge and superstition to systematic observation, identification of causal agents, and the beginnings of a scientific approach to understanding and managing plant diseases. These developments set the stage for the more advanced research and practices that would follow in the 20th and 21st centuries.

Discovery of the Potato Late Blight (Mid-19th Century)

The Irish Potato Famine of the mid-19th century was a major turning point in plant pathology. It was caused by the oomycete pathogen *Phytophthora infestans*. This event prompted further scientific investigation into plant diseases.

The discovery of the potato late blight in the mid-19th century stands as a pivotal moment in agricultural history, marked by profound consequences and scientific enlightenment. This devastating plant disease, caused by the oomycete pathogen *Phytophthora infestans*, emerged in the early 1840s and quickly escalated into a catastrophic event. The disease manifested as dark, irregularly shaped lesions on the leaves and tubers of potato plants, rendering them inedible. The Irish Potato Famine, resulting from this blight, had far-reaching humanitarian and socio-economic ramifications, including mass emigration, population decline, and fundamental changes in Irish society.

However, the tragic consequences of the potato late blight also prompted a renewed commitment to understanding plant diseases and drove scientific investigation. Naturalists and botanists began to systematically study the disease, documenting its symptoms and spread, leading to pioneering insights into its causal agent. The work of botanist Miles Joseph Berkeley contributed to the recognition of the late blight as a disease caused by a microscopic organism, although the true nature of the oomycete pathogen was not yet fully understood.

The discovery of the potato late blight laid the groundwork for modern plant pathology, as scientists subsequently pursued a more scientific understanding of plant diseases, including their causes and effective management strategies. This tragic chapter in agricultural history served as a catalyst for the systematic study of plant diseases, ultimately contributing to more comprehensive knowledge and strategies for disease prevention and management in the years to come.

The discovery of the potato late blight in the mid-19th century is a pivotal moment in the history of plant pathology. It not only had profound consequences for agriculture but also served as a catalyst for the systematic study of plant diseases.

Background

The potato, a staple crop in Ireland, had become a primary source of sustenance for a significant portion of the population by the mid-19th century.

The pathogen responsible for the late blight of the potato was later identified as *Phytophthora infestans*, an oomycete microorganism. However, its true nature was not understood at the time.

Key Events

- **Emergence of the Disease**: In the early 1840s, a devastating plant disease began to afflict potato crops in Ireland. The disease manifested as dark, irregularly shaped lesions on the leaves, which quickly spread to the tubers. The infected potatoes rotted, rendering them inedible.
- **Rapid Spread**: The disease's rapid spread and severity led to catastrophic consequences. Potato crops across Ireland were decimated, leading to widespread food shortages and a severe humanitarian crisis.

- **Efforts to Combat the Disease**: Farmers and agricultural authorities made various attempts to combat the disease, including crop rotation and destroying infected plants, but these measures were largely ineffective in halting the spread of the disease.
- **Scientific Investigation**: The devastating impact of the potato late blight caught the attention of scientists and naturalists. They began to systematically investigate the disease, documenting its symptoms and spread.
- **Contributions of Scientists**: Notably, the work of the botanist Miles Joseph Berkeley was influential. He recognized that the late blight was due to a microscopic organism but was unable to identify it precisely as an oomycete.
- **Legacy and Impact**: The Irish Potato Famine, caused by the late blight, had profound and lasting effects. It led to mass emigration, population decline, and socio-economic changes in Ireland. Additionally, it underscored the need for scientific inquiry into plant diseases and set the stage for more systematic research in plant pathology.
- **Advancements in Plant Pathology**: The study of the potato late blight played a pivotal role in the development of plant pathology as a scientific discipline. Subsequent researchers, such as Anton de Bary, would further explore the nature of plant diseases and lay the foundation for understanding their causes and control.

The discovery of the potato late blight in the mid-19th century was a tragic chapter in agricultural history, but it had a silver lining in that it prompted scientists to delve into the systematic study of plant diseases, eventually leading to a more comprehensive understanding of these afflictions and the development of methods for their control.

The Germ Theory of Disease (Late 19th Century)

The work of Louis Pasteur and Robert Koch on the germ theory of disease, which applied to both humans and plants, greatly advanced the understanding of infectious agents and their role in causing diseases. The late 19th century marked a transformative era in the history of medicine and biology with the advent of the germ theory of disease. This groundbreaking theory, which fundamentally altered humanity's understanding of the causes of illness, proposed that many diseases, both in humans and plants, were not merely the result of vague, mystical forces or miasmas, but were instead caused by microscopic organisms – bacteria and other microbes. The germ theory of disease represented a paradigm shift that revolutionized medical and plant pathology, leading to significant advances in disease diagnosis, prevention, and treatment. This pivotal concept not only laid the groundwork for modern medicine but also played a crucial role in shaping the field of plant pathology by extending our comprehension of the causes of plant diseases and revolutionizing approaches to their management. In this context, we delve into the late 19th century, a time when scientists, armed with microscopes and a thirst for knowledge, made remarkable strides in uncovering the invisible world of disease-causing germs, reshaping the course of human and plant health for generations to come.

The late 19th century marked a transformative era in the history of medicine and biology, characterized by the revolutionary advent of the germ theory of disease. This groundbreaking scientific concept, which forever altered humanity's understanding of the causes of illness, proposed that many diseases, both in humans and plants, were not merely the result of vague, mystical forces or miasmas, but were instead caused by invisible, microscopic organisms – bacteria and other microbes.

Key Developments and Highlights of the Germ Theory of Disease in the Late 19th Century

- **The Work of Louis Pasteur:** French scientist Louis Pasteur is often regarded as one of the pioneers of the germ theory. In the 1860s, Pasteur conducted a series of experiments that demonstrated the role of microorganisms in fermentation and the spoilage of food and beverages. His work provided compelling evidence for the existence of microscopic life forms as agents of decay and disease.
- **Robert Koch's Postulates:** German physician and microbiologist Robert Koch introduced a set of postulates in the 1880s, known as Koch's postulates, which provided a framework for definitively linking specific microbes to particular diseases. These postulates laid the foundation for the systematic study of infectious diseases.
- **Advances in Microscopy:** The development of improved microscopes allowed scientists to observe and study microorganisms in unprecedented detail. This technological advancement was pivotal in confirming the presence of bacteria and other microbes in disease processes.
- **Applications in Medicine:** The germ theory of disease greatly influenced medical practice. Physicians began to adopt aseptic techniques, such as sterilization, to reduce the spread of infections during surgeries and medical procedures. This dramatically improved patient outcomes.
- **Impact on Plant Pathology:** The principles of the germ theory were not confined to human medicine. They also had a profound impact on the field of plant pathology. Plant pathologists recognized that many plant diseases were caused by microorganisms, including bacteria and fungi. This insight led to the development of methods to diagnose, prevent, and manage plant diseases.

- **Public Health and Hygiene:** The germ theory revolutionized public health practices. It emphasized the importance of sanitation, clean water, and hygiene in preventing the spread of diseases. This led to significant improvements in public health and a reduction in the incidence of infectious diseases.
- **Vaccination and Immunization:** Building on the germ theory, researchers such as Edward Jenner and Louis Pasteur developed vaccines against diseases like smallpox and rabies. Vaccination became a cornerstone of preventive medicine.

The late 19th-century germ theory of disease represented a profound shift in the way society perceived and approached health and disease. It not only laid the foundation for modern medicine but also significantly advanced the field of plant pathology by expanding our understanding of the causes of plant diseases and revolutionizing strategies for their management. The germ theory's legacy endures to this day, underpinning our approach to both human and plant health and emphasizing the critical role of microorganisms in the natural world.

Emergence of Bacteriology and Virology (Late 19th to Early 20th Century)

The late 19th and early 20th centuries saw significant progress in the study of plant pathogens. Bacteriologists and virologists identified bacteria and viruses as agents responsible for plant diseases. This period also marked the development of methods for diagnosing and controlling plant diseases. The late 19th to early 20th century marked a remarkable era of scientific exploration and discovery, particularly in the fields of bacteriology and virology. This period witnessed the emergence of two distinct but interrelated disciplines that would forever transform our understanding of infectious agents, human and animal health, and agriculture. Bacteriology, the study of bacteria, and virology, the study of viruses, brought to light the existence of microscopic entities responsible for a multitude of diseases in humans, animals, and plants. These groundbreaking discoveries not only revolutionized medical and agricultural practices but also paved the way for the development of vaccines, antibiotics, and a deeper understanding of the complex interactions between pathogens and their hosts. This narrative explores the key milestones and pioneers in the fields of bacteriology and virology during this period, shedding light on the transformative impact of these disciplines on science and society.

The late 19th to early 20th century witnessed a profound transformation in the fields of science and medicine with the emergence of bacteriology and virology. This pivotal period was marked by groundbreaking discoveries, innovative research methodologies, and the establishment of two distinct but interconnected disciplines that forever altered our comprehension of infectious agents, human and animal health, and agricultural practices.

Key Developments and Highlights of the Emergence of Bacteriology and Virology

- **Bacteriology Takes Root:** The late 19th century saw the establishment of bacteriology as a formal scientific discipline. Pioneers such as Louis Pasteur, Robert Koch, and Ferdinand Cohn made groundbreaking contributions. Louis Pasteur's work on fermentation and the role of microorganisms in spoilage laid the foundation for understanding bacteria's significance in both disease and industry.
- **Robert Koch's Postulates:** German physician and microbiologist Robert Koch introduced Koch's postulates, a set of criteria that provided a systematic approach to proving that a particular microorganism causes a specific disease. His work on anthrax and tuberculosis established these pathogens as bacterial diseases.
- **Development of Microscopy:** Advancements in microscopy, notably the development of high-powered lenses, allowed scientists to observe bacteria with unprecedented clarity. This technological progress was integral to confirming the role of bacteria in infectious diseases.
- **Virology Takes Shape:** The concept of viruses as infectious agents began to emerge. Dmitry Ivanovsky and Martinus Beijerinck independently conducted experiments with infectious agents that could pass through bacterial filters but were smaller than bacteria. These agents, termed "filterable viruses," marked the beginning of virology.
- **Discovery of Bacteriophages:** Frederick Twort and Félix d'Hérelle discovered bacteriophages, viruses that infect and replicate within bacteria. This discovery provided further evidence of the existence of viruses and their potential role in diseases.
- **Vaccination and Immunization:** Building on these developments, vaccines against bacterial diseases, such as diphtheria, pertussis, and tetanus, were developed. The field of immunization advanced significantly, leading to the prevention of numerous diseases.
- **Antibiotics:** The discovery of antibiotics, such as penicillin by Alexander Fleming, revolutionized medicine. Antibiotics offered effective treatments for bacterial infections and significantly improved patient outcomes.
- **Impact on Agriculture:** Bacteriology and virology also played a crucial role in agriculture. Plant pathologists recognized that bacteria and viruses were responsible for many plant diseases. This understanding led to the development of strategies to diagnose, control, and prevent crop diseases.

The late 19th to early 20th century marked a turning point in scientific and medical history, as bacteriology and virology emerged as distinct disciplines. These fields brought to light the role of microorganisms in causing diseases and laid the foundation for modern medicine, microbiology, and agricultural science. The groundbreaking discoveries made during this era continue to shape our approach to health, disease prevention, and food production in the 21st century.

The Green Revolution (Mid-20th Century)

The mid-20th century saw a significant shift towards understanding and controlling plant diseases as part of the Green Revolution. This period was characterized by the development and widespread adoption of disease-resistant crop varieties and the use of fungicides and pesticides to protect crops. The mid-20th century witnessed a monumental transformation in agriculture and food production known as the Green Revolution. This unprecedented period of innovation and scientific advancement brought about a radical shift in the way the world cultivated crops and addressed the pressing issue of global food security. Rooted in advancements in crop breeding, pest and disease management, and the widespread adoption of modern agricultural practices, the Green Revolution ushered in an era of remarkable agricultural productivity. By introducing high-yielding crop varieties, improved irrigation methods, and the use of synthetic fertilizers and pesticides, it not only significantly increased crop yields but also played a pivotal role in alleviating famine and hunger across the world. This narrative explores the key elements and luminaries of the Green Revolution, underscoring its far-reaching impact on global agriculture, economics, and food availability.

The mid-20th century marked a transformative period in agriculture and global food production known as the Green Revolution. This unprecedented era of innovation and scientific progress brought about a profound shift in the way the world approached crop cultivation and addressed the urgent challenge of feeding a rapidly growing global population.

Key Developments and Highlights of the Green Revolution

- **High-Yielding Crop Varieties:** One of the cornerstones of the Green Revolution was the development and widespread adoption of high-yielding crop varieties. Scientists such as Norman Borlaug, often considered the father of the Green Revolution, bred wheat varieties with significantly increased yields. These new cultivars were instrumental in boosting food production.
- **Modern Agricultural Practices:** The Green Revolution introduced modern agricultural practices, including the use of synthetic fertilizers, improved irrigation methods, and the widespread application of pesticides and herbicides. These practices optimized crop growth and minimized losses due to pests and diseases.

- ☆ **Global Impact:** The Green Revolution was not confined to a single region. It had a global reach, with initiatives spanning across Asia, Latin America, and Africa. India and Mexico were among the early adopters, and the Green Revolution's principles were extended to numerous crops.
- ☆ **Economic Transformation:** Increased agricultural productivity had profound economic effects. It led to higher incomes for farmers, reduced rural poverty, and helped shift economies from agrarian to industrial and service-oriented.
- ☆ **Food Security:** By significantly increasing crop yields, the Green Revolution played a pivotal role in addressing food security concerns. It helped ensure a more stable and adequate food supply for growing populations.
- ☆ **Critiques and Challenges:** While the Green Revolution brought remarkable gains in food production, it was not without challenges. Critics pointed to environmental concerns, such as soil degradation and water resource management. There were also concerns about the potential loss of crop diversity.
- ☆ **Long-Term Sustainability:** Efforts have been made to ensure the long-term sustainability of the Green Revolution's gains. Sustainable agriculture practices, conservation of biodiversity, and organic farming approaches have been incorporated to address some of the challenges.
- ☆ **Technological Advancements:** The Green Revolution leveraged significant technological advancements in agriculture, such as the development of hybrid crops, better understanding of plant genetics, and improved farm machinery.

The Green Revolution remains a defining moment in the history of agriculture and food production. It not only helped prevent widespread famine and hunger but also catalyzed economic development and propelled the world towards the complex challenges and opportunities of modern agriculture. As global food needs continue to evolve, the lessons and innovations of the Green Revolution remain a valuable source of knowledge in our ongoing quest for sustainable and secure food production.

Biotechnology and Genetic Engineering (Late 20th Century)

Advances in biotechnology, including the development of genetically modified (GM) crops with resistance to pests and diseases, played a pivotal role in managing plant diseases. The late 20th century marked a watershed moment in the history of science and agriculture with the advent of biotechnology and genetic engineering. This transformative era witnessed a revolution in our ability to manipulate the genetic material of organisms, unlocking unprecedented potential for advancing medicine, industry, and agriculture. At the heart of this scientific revolution was the capacity to splice and modify DNA, which not only led to the development of life-saving pharmaceuticals but also ignited a paradigm shift in crop improvement.

The application of genetic engineering techniques to agriculture ushered in a new age of crop breeding, enabling the creation of genetically modified organisms (GMOs) with desired traits, enhanced nutritional value, and increased resistance to pests and diseases. This narrative explores the profound impact of biotechnology and genetic engineering on agriculture, highlighting their contributions to crop yield, food security, and sustainable farming practices, while also examining the ethical, environmental, and socio-economic considerations that continue to shape this dynamic field.

The late 20th century witnessed a monumental and revolutionary transformation in science and agriculture with the advent of biotechnology and genetic engineering. This transformative era marked a shift in our capacity to manipulate the fundamental building blocks of life, DNA, and unlock unprecedented potential in medicine, industry, and agriculture.

Key Developments and Highlights of Biotechnology and Genetic Engineering in the Late 20th Century

- **Emergence of Recombinant DNA Technology:** Recombinant DNA technology, also known as genetic engineering, allowed scientists to manipulate the genetic material of organisms with precision. This marked a pivotal moment in the history of science, offering the power to splice and modify genes for specific purposes.
- **Creation of Genetically Modified Organisms (GMOs):** Genetic engineering was applied to agriculture, leading to the creation of genetically modified organisms (GMOs). These organisms, including crops, were engineered to possess desirable traits, such as resistance to pests, tolerance to herbicides, and improved nutritional content.
- **Crop Improvement:** Genetic engineering enabled the development of crop varieties with increased yield, reduced susceptibility to diseases, and enhanced nutritional value. Crops like Bt cotton and genetically modified maize became emblematic of the potential benefits of biotechnology in agriculture.
- **Pest and Disease Resistance:** Genetic engineering led to the creation of crops resistant to pests and diseases, reducing the need for chemical pesticides and lowering the environmental impact of agriculture.
- **Food Security:** Biotechnology and genetic engineering played a critical role in addressing global food security concerns by increasing crop yields and reducing post-harvest losses. They helped ensure a more stable and adequate food supply for a growing global population.

- ☆ **Challenges and Controversies:** The widespread adoption of GMOs also raised concerns and controversies. Environmental, safety, and labeling issues generated public debate and regulatory challenges.
- ☆ **Medicine and Pharmaceuticals:** Beyond agriculture, biotechnology had a profound impact on medicine and pharmaceuticals. It led to the development of life-saving drugs, such as insulin, growth hormones, and monoclonal antibodies, as well as breakthroughs in gene therapy.
- ☆ **Ethical and Socio-economic Considerations:** The advent of biotechnology raised ethical questions about the modification of organisms and the ownership of genetic material. Socio-economic disparities in access to and benefits from biotechnology were also highlighted.
- ☆ **Regulatory Frameworks:** Governments and international bodies developed regulatory frameworks to oversee the application of biotechnology in agriculture and medicine, ensuring safety and addressing public concerns.

The late 20th-century biotechnology and genetic engineering revolutionized agriculture, offering the potential to address food security and sustainability challenges. These advancements extended far beyond farming, profoundly impacting medicine and industry. Yet, the field also encountered ethical, environmental, and regulatory challenges that continue to shape the ongoing discourse on biotechnology's role in shaping our future.

Sustainable Agriculture (21st Century)

The 21st century has witnessed a growing emphasis on sustainable agriculture and integrated disease management. Research focuses on environmentally friendly methods, biological control, and the reduction of chemical pesticide use.

The 21st century has ushered in a new era in agriculture, one defined by a resounding commitment to sustainable practices and environmental stewardship. As the world grapples with the complex challenges of a growing global population, climate change, and the preservation of our planet's ecosystems, sustainable agriculture has emerged as a transformative force. This approach to farming focuses on the harmonious coexistence of agriculture and the environment, seeking to maximize productivity while minimizing negative impacts on the planet.

Sustainable agriculture encompasses a broad spectrum of principles and practices, including organic farming, crop rotation, reduced pesticide use, water conservation, and the promotion of biodiversity. It champions responsible resource management and strives to create resilient and regenerative agricultural systems. Sustainability in agriculture not only benefits the environment but also promotes economic viability for farmers and ensures the availability of safe, nutritious, and accessible food for people worldwide. It stands as a testament to humanity's collective commitment to addressing the pressing issues of food security, climate change, and the protection of our natural world, forging a path towards a more sustainable and secure future.

The 21st century has marked a profound shift in agriculture, with a growing emphasis on sustainable practices that prioritize environmental conservation, social equity, and economic viability. Sustainable agriculture, often seen as a response to the challenges of a rapidly changing world, has become a driving force in reshaping the way we produce food and manage our natural resources.

Key Developments and Highlights of Sustainable Agriculture in the 21st Century

- ☆ **Environmental Stewardship:** Sustainable agriculture places a strong emphasis on environmental sustainability. It seeks to reduce the ecological footprint of farming by promoting practices that minimize soil erosion, preserve biodiversity, and conserve water resources.
- ☆ **Organic Farming:** The organic farming movement has gained momentum in the 21st century, with more farmers adopting organic practices that avoid synthetic pesticides and fertilizers while emphasizing soil health and biological diversity.
- ☆ **Crop Diversity and Rotation:** Crop rotation and the cultivation of a diverse range of crops in the same field have gained popularity. These practices reduce soil degradation, minimize disease and pest pressure, and enhance overall soil fertility.
- ☆ **Conservation Agriculture:** Conservation agriculture techniques, such as minimal or no-tillage farming, have been widely adopted to reduce soil disturbance, retain soil moisture, and increase carbon sequestration in the soil.
- ☆ **Water Management:** Sustainable agriculture focuses on efficient water management through techniques like drip irrigation, rainwater harvesting, and the use of drought-resistant crop varieties, addressing the growing issue of water scarcity.
- ☆ **Integrated Pest Management:** Integrated pest management (IPM) approaches aim to reduce pesticide use by combining biological, cultural, and chemical control methods, helping to protect beneficial insects and pollinators.
- ☆ **Local and Small-Scale Farming:** The support for local and small-scale farming has grown, fostering community-based agriculture and reducing the environmental footprint associated with long-distance food transportation.

- ☆ **Regenerative Agriculture:** Regenerative agriculture practices seek to restore and improve soil health, with a focus on carbon sequestration, increasing farm resilience, and enhancing ecosystem services.
- ☆ **Food Security and Nutrition:** Sustainable agriculture plays a critical role in addressing global food security and improving access to nutritious food for all. Initiatives like urban farming and community gardens contribute to local food production.
- ☆ **Global Initiatives:** International organizations, governments, and agricultural institutions have recognized the importance of sustainability in agriculture, setting goals and guidelines to promote responsible farming practices.
- ☆ **Climate Resilience:** As climate change poses new challenges for agriculture, sustainable practices contribute to climate resilience by reducing greenhouse gas emissions and adapting to changing weather patterns.
- ☆ **Consumer Awareness:** Growing consumer awareness and demand for sustainably produced food have driven changes in agricultural practices and supply chains.

Sustainable agriculture in the 21st century represents a holistic and forward-looking approach to food production, recognizing that the well-being of people, the planet, and future generations are interconnected. It exemplifies a shared commitment to addressing the environmental and social challenges that shape the future of agriculture, making it a cornerstone of a more resilient and sustainable global food system.

Emerging Threats (Ongoing)

Climate change and global trade have brought about new challenges for plant pathology. The emergence of new pathogens, pests, and diseases highlights the ongoing need for research and strategies to protect global food security.

Throughout its history, plant pathology has evolved from observation and mysticism to a highly specialized and scientific field. Plant pathologists play a crucial role in understanding, preventing, and managing diseases that impact crops, which is vital for ensuring global food production and agricultural sustainability.

In the 21st century, agriculture faces a dynamic landscape characterized by ongoing and emerging threats that challenge the sustainability, productivity, and resilience of our food systems. These threats are multi-faceted and include the increasing impacts of climate change, the degradation of natural resources, the spread of new and existing pests and diseases, and the complex interplay of global factors such as trade and socio-economic dynamics. Climate change, with its unpredictable weather patterns, extreme events, and altered growing conditions, poses a profound threat to agricultural stability, necessitating adaptive strategies and resilient crop varieties. The degradation of soils, depletion of water resources, and loss of biodiversity further undermine the long-term sustainability of farming. In addition, evolving pests and diseases, often exacerbated by global trade and the movement of people and goods, require ongoing vigilance and innovative solutions. The interconnectivity of these threats underscores the need for holistic, science-based approaches that not only address the immediate challenges but also build towards a more sustainable, equitable, and food-secure future. As the agricultural community continues to adapt and innovate, these ongoing and emerging threats serve as a powerful reminder of the collective effort required to safeguard the global food supply and ensure the well-being of people, communities, and the environment in the face of evolving challenges.

In the 21st century, agriculture faces a continually evolving landscape of emerging threats that demand attention and innovative solutions. These challenges encompass a wide spectrum of factors that impact food production, environmental sustainability, and global food security.

Key Ongoing and Emerging Threats in Agriculture

- **Climate Change:** The ongoing impacts of climate change, including shifting weather patterns, increased temperatures, and extreme events, threaten crop yields, water availability, and the overall stability of agricultural systems. Adaptation and mitigation measures are essential.
- **Pest and Disease Evolution:** Pests and pathogens continually evolve, posing new threats to crops and livestock. The global movement of goods and people accelerates the spread of these threats, requiring vigilance and proactive management.
- **Soil Degradation:** Soil erosion, nutrient depletion, and soil degradation are ongoing concerns that affect crop productivity. Sustainable soil management practices are crucial to address these challenges.
- **Biodiversity Loss:** The loss of biodiversity impacts ecosystem services that are essential for agriculture, such as pollination, pest control, and soil health. Conserving and restoring biodiversity is a priority.
- **Water Scarcity:** Ongoing water scarcity issues, exacerbated by over-extraction and pollution, challenge irrigation-dependent agriculture. Efficient water management practices are essential to mitigate this threat.
- **Food Safety and Quality:** As food systems become more complex and globalized, ensuring food safety and quality remains an ongoing challenge. Contaminants, foodborne pathogens, and adulteration require robust monitoring and regulatory systems.

- **Global Trade Dynamics:** Trade policies, tariffs, and geopolitical factors can disrupt global food supply chains and have significant economic and agricultural consequences.
- **Consumer Preferences and Demands:** Evolving consumer preferences for healthier and more sustainably produced food products are changing the agricultural landscape and influencing production practices.
- **Resource Scarcity:** Scarcity of critical resources, including land and energy, poses challenges to agriculture. Efficient resource use and diversification are important responses.
- **Technology and Data Security:** The increasing reliance on digital technologies and data in agriculture introduces new threats related to data security, privacy, and cyber-attacks.
- **Social and Economic Inequalities:** Socio-economic disparities and inequities in access to resources, knowledge, and opportunities in agriculture persist and impact the sector's sustainability.
- **Political and Regulatory Changes:** Shifting political landscapes and regulatory changes can influence agricultural policies, trade agreements, and research funding, affecting the sector's trajectory.
- **Invasive Species:** The introduction and spread of invasive species, whether plant pests, animal diseases, or aquatic invaders, pose a persistent threat to agricultural ecosystems and biodiversity.

Addressing these ongoing and emerging threats in agriculture requires a multi-faceted approach that combines scientific innovation, policy development, sustainable practices, and international cooperation. As agriculture adapts to this dynamic landscape, it must strive for resilience, sustainability, and equitable access to resources, all while producing the nutritious and safe food needed to nourish a growing global population.

Chapter 3

Indian Plant Pathologist

The science of plant pathology in modern period in this country is rather not very old. Even when humans lived as hunters or nomads and their food consisted only of meat or leaves, fruit, and seeds. Which they picked wherever they could find them, plant diseases took their toll on hunted animals and on humans. The development of this subject started with the study of mycology.

The first study of pathology started in Indian universities that were established in 1857 at Calcutta, Bombay and Madras emphasized collection, identification, characterization and taxonomy of fungi. Up to 1930, there was more attention to the study of fungi, bacteria, virus, MLOs and Nematode than to disease caused by them.

Many scientists are contributed in plant pathology are mentioned here such as:

D.D. Cunnigham (1850-1875)

Started identification of fungi in this country.

- ☆ Cunnigham was found a species *Gymnosporangium* fungus, and named *G. cunninghamianum*.
- ☆ His studied the spread of bacteria and the spores of fungi through the air and also made studies on cholera.
- ☆ Cunningham made a special study of rusts and smuts.
- ☆ Worked and collected the members of mucorales and uredinales.

- ☆ Identified the causal organism of red rust of tea in Assam caused by *Cephaleurous virescens.*

A. Barclay

- ☆ Barclay began his critical work on the rust near Shimla, and during his following 7 years.
- ☆ He studies the life histories of many of the Himalayan Uredinales.

K.R. Kirtikar

- ☆ Kirtikar was the first Indian scientist who collected and identified fungi.
- ☆ Kirtikar was collect and identified some of mushrooms. He wrote a Book- Indian Medical Plants

E.J. Butler (1874-1943)

- ☆ He was the father of modern plant pathology and father of Indian Mycology.
- ☆ He worked at IARI for 20 year from 1901 to 1920.
- ☆ Butler stayed in India for about 20 year and besides training several Indian scientists, studies wilt of cotton and pigeon pea, disease of rice, toddy palm, sugarcane, potato and cereal rust.
- ☆ He was the founder and first director of Imperial Mycological Institute, England (1920-35).
- ☆ Monograph: *Pythiaceous* and allied *fungi.*
- ☆ He wrote a Book: Fungi and Disease in plants (1918).

B.B. Mundkur

- ☆ He worked as Asst. Mycologist in the cotton research scheme at Dharwar, and after appointed Asst. Mycologist in (IARI), Pusa, Bihar.
- ☆ He worked on the control of cotton wilt disease through varietal resistance.
- ☆ He was also responsible for the identification and Indian classification of large number of smut fungi.

- ☆ He started Indian Phytopathological Society in 1948 with its published a ***Journal Indian Phytopathology*** published about 80 papers.
- ☆ He also authored a textbook: **Fungi and plant diseases,** and the monographs 'Genera **of Rusts'.**

J.F. Dastur (1886-1971)

- ☆ Born on 10 sep. 1886 at Bhavnagar; M.Sc. in 1909 Bombay university, Entered government service in 1910 as a first Asst. to the Imperial Mycologist and appointed as Imperial Mycologist and Head of Plant Pathology, IARI, New Delhi.
- ☆ Dastur a colleague of Butler, was the first Indian plant pathologist. Who is credited with the detailed study of fungi and plant diseases? His special field of study was the genus *Phytophthora* and *Corticium* diseases of castor and potato caused by its species.
- ☆ He established *Phytophthora parasitica* from castor. He has published about 36 original papers and memoirs

S.L. Ajrekar

- ☆ Ajrekar was a professor of botany in Institute of science, Mumbai (1923-28) and science institute, Gujarat College, Ahmedabad (1928 38). Ajrekar studied wilt of cotton, *Mucor, Aspergilus,* smut of sugarcane, and ergot of sorghum.
- ☆ Ajrekar was an Editor of Journal of the Indian Botanical Society.
- ☆ The Maharashtra Association for the cultivation of Science named its herbarium 'Ajrekar Mycological Herbarium' in his honour.

K.D. Bagchee (1898-1973)

- ☆ Born in January 1898. Obtained M.Sc. Degree in Botany from Calcutta University and D.Sc. and D.I.C. from London University.
- ☆ Started career as Mycologist in 1927 at the Forest Research Institute, Dehradun and did pioneering research on mycology and Forest Pathology.

- ✰ Special reference to biology of rust fungi on conifers, soil borne diseases and their control.
- ✰ After retirement in 1956 from the (F.R.I), appointed Director, Botanical Research, J&K State where he initiated and organized work on botanical research and surveyed diseases of forest plants in the state.

Dr. R. Prasad (1907-94)

- ✰ Dr. Prasada was born on 11 Dec., 1907 at Chandausi, in (U.P). He was educated at Agra College, taking B.Sc., M.Sc. and D.Sc. Degrees of Agra University.
- ✰ In 1946 Dr. Prasad joined the IARI, New Delhi and initiated work on the epidemiology of cereal rust in association with Dr. K. C. Mehta.
- ✰ He had also been associated since 1935 with the breeding of rust resistant varieties of wheat.
- ✰ Dr. Prasad has been closely connected with the Indian Phytopathological Society. He was editing the journal Indian Phytopathology for six year as Chief-Editor.

S.N. Das Gupta (1902-90)

- ✰ Born at Jalpaigury on 4th Nov. 1902 Educated at Bangabasi College, presidency college Calcutta passed M.Sc. (Botany).He awarded Ph.D. and D.Sc. Degrees and D.I.C. of London University.
- ✰ He started his career as Reader in Botany at the University of Lucknow (1934) and became Professor and (H.O.D) Botany, in 1949, and Dean, faculty of science in 1954.
- ✰ He developed a school of Plant Pathology including virology and microbiology at Lucknow University.
- ✰ Conducted researches in sanitation of fungi, nutritional requirement of fungi, mango necrosis, and deficiency disease of mango, aquatic fungi and medical mycology.
- ✰ Dasgupta studies on the controversial mango black tip disease.
- ✰ Dasgupta studied the role of enzymes in pathogencity

M.J. Thirumalachar (1914-99)

- ☆ Born in sep. 1914 at Bangalore, received the B.Sc. Botany, and D.Sc. from the University of Mysore; worked under the guidance of Dr. B.B. Mundkur.
- ☆ He created over 20 new genera and 300 new species of fungi.
- ☆ Thirumalachar conducted exhaustive study of smuts, rusts and *Cercospora etc.* He studied their taxonomy, cytology and life-cycles.
- ☆ He successfully introduced the use of antibiotics in plant disease control. Ex. Aureofungin and streptocycline.
- ☆ While in USA, he isolated a new genus, *Geotrichum marinum* which can grow on the oil, carbon and nitrogen source?

Prof. M.K. Patel (1899-1967)

- ☆ Prof. Patel laid the foundation of phytobacteriology in India.
- ☆ He established a school of Plant Bacteriology at college of agriculture, Pune and first described a new sp. *Xanthomonos campestris pv. uppali* in 1948, from the host *Ipomea muricota.*
- ☆ He first documented that black spot of mango is caused by *Pseudomonas mangiferaeindiceae.*
- ☆ He described more than 30 bacterial diseases from India such as- Bacterial leaf spot on *Vernonia cinerea* in 1968. Bacterial leaf spot of *Cyamopsis tetragonoloba* in 1953.17

Abrar M. Khan (1919-2004)

- ☆ He was prof. and Head, Department of Botany at (A.M.U), Aligarh. He obtained a Ph.D. with Prof. Stakman.
- ☆ He worked the development of *Sphareo-theca* and *Erisiphe* infecting cucurbits in tropical climate.
- ☆ In India, he pioneered research on phytonematology. He was also considered as the father of Nematology in India and initiated the Nematological Society of India.

Dr. H.K. Saksena (1922-2007)

- ✰ Born on 21 Dec. 1922 at Sultanpur; M.Sc. in Botany, (AU) 1944; Associate IARI in mycology and plant pathology.
- ✰ Started his career in 1952 as lecturer in Botany, (AU) In 1953.
- ✰ Joined as Assistant Prof. of Plant Pathology and continued as Prof. and Head of Deptt. of Plant Pathology since1961 at Govt. Agri. College of Kanpur.
- ✰ Associated with the Indian Phytopathological Society since its inception as a Charter Member, Councilor, Vice President, President and Editor in Chief.
- ✰ His main contribution have been on the parasitism and saprophytic survival of plant pathogenic fungi, and on the taxonomy, cytology, and physiology of parasitism.
- ✰ His main study on *R. solani* and control of *Rhizoctonia* group of fungi.

Prof. R.S. Singh (1927)

- ✰ Dr. Singh served as Prof. and Head Deptt. of Plant Pathology at Pantnagar.
- ✰ He worked on biological control and ecology of soil borne pathogen.
- ✰ His book:- "Plant diseases" (1963) is referred as Bible of Plant Pathology in India.

  Principles of Plant Pathology, Diseases of Vegetable Crops, Diseases of Fruit Crops, Plant Disease Management and A Handbook of Economic Nematology'.

Prof. R.K. Grover (1931-2007)

- ✰ He worked mainly at Haryana Agriculture University, Hisar, on fungicidal control of plant diseases specially caused by *Rhizoctonia* diseases of crop plants.
- ✰ He worked Chemical control of root rot of cow pea and Studies on *Rhizobium*-fungitoxicant interaction in cow pea.
- ✰ He introduced the concept of environmental factors of soil and aerial parts on the effectiveness of fungicides.

Dr. B.B. Nagaich

- ✰ Dr. Banke Bihari Nagaich was born 11th Aug. 1931 at District Hamirpur (U. P.). In the year 1955 are obtained M.Sc. (Ag.) Degree in Plant Pathology from Agra University.

- Later he obtained his Doctral Degree from the University of Illionis (US). During his stay at Illionis he was assisted by Prof. Thornberry in survey and surveillance work on plant viruses.
- Dr. Nagaich joined the IARI at its Regional Station at Shimla in 1958.
- He joined the CPRI, Shimla where he worked as a virologist and as Head, Division of Plant Pathology.
- In Oct.1975, Dr. Nagaich took over as the Director of CPRI, Shimla(1975-82) and retired from this post in 1982.
- In 1980-81, he was the first Editor-in –Chief of the journal of *Indian Potato Association.*
- Dr. Nagaich is also the life member of Indian Phytopathological Society. He was the President of IPS in 1982.

Dr. Y.L. Nene (1936)

- Yeshwant Laxman Nene was born in Gwalior, Madhya Pradesh in 1936.
- He did his undergraduate education in Agriculture College at Gwalior and post-graduate education in plant pathology at the Agricultural College, Kanpur.
- He started his professional career in 1960 as Assistant Professor and the founder of the Department of Plant Pathology, (GBPUA&T) Pantnagar, U.K., and India.
- In 1989, he was appointed the Deputy Director General of ICRISAT, Hyderabad.
- Dr. Nene, while working at GBPUA&T, solved the mystery of **khaira disease of rice** and established that it was caused by zinc deficiency.
- At ICRISAT, he developed multiple-disease resistance-screening techniques and disease-resistant varieties of pulses.
- Authored the books ***"Fungicides in plant disease control"*** and "***The pigeonpea***".
- It was on 30 Dec. 1994 that he established the AAHF and a Trust, (AAHFT). He was elected as the first Chairman of the AAHFT.

Dr. B.S. Mehrotra

- ☆ Born in Allahabad in 1927, obtained his M.Sc. Degree, Botany Deptt., (AU).
- ☆ He did his Ph.D. under the guidance of Dr. R.K. Saksena from (A.U).
- ☆ He made excellent contributions on taxonomy of fungi particularly of *Aspergilli* and *Penicillia.*
- ☆ He retired from Head, Department of Botany, (AU).
- ☆ He has written a book **'The Fungi'.**

A.N. Mukhopadhyay (1940)

- ☆ He did his Ph.D. from B.H.U., He served as professor. and Head Deptt. of Plant Pathology at Gujarat Agriculture University in 1982.
- ☆ He is known for his work on bio-control of plant disease in Pantnagar University (1984).He is credited with the publication of four edited volumes entitled with ***Plant Disease of International Importance***

C. Manoharachary (1944)

- ☆ He was Prof. of Botany, Osmania University, Hyderabad, and now Emeritus professor there.
- ☆ The research conducted in bio-diversity of fungi, fungal 'bio-technology, bio-pesticides, plant protection, bio-fertilizer technology, Mushroom cultivation and advances areas received recognization at National and International level.
- ☆ He contributed extensively on the various aspects of soil fungi,Aquatic fungi, forest fungi.
- ☆ Executive editor, Indian Journal of Microbial Ecology since 1991.
- ☆ He has been awarded "**E.K. Janaki Ammal National Award"** for taxonomy by the Ministry of Environment and Forestry,Government of India on 10 Oct. 2006.
- ☆ Research publication-400National -320International - 80Books - 12
- ☆ Author a Monograph "Soil Fungi from Andhra Pradesh"(1990)
- ☆ Author Text Book:- "Botany-Lower Plants"(1993), "Fungi"(1995), "Botany"(1996), "Cultivation of Mushrooms"(1996).

Dr. Anandraj

- ✰ Born in Karnataka, India. Completed M.Sc. From University of Mysore, and Ph.D. from University of Calicut.
- ✰ He was Senior Scientist (1984-1998), Principal Scientist, (1998-2005), Head, Division of Crop Protection, (IISR) Calicut, Kerala, (2005-2006).
- ✰ Director, Indian Institute of Spices Research, Calicut (July, 2011 onwards). His research worked on Plant Pathology- Epidemiology, Soil borne diseases especially *Phytophthora*, Biological Control.

Chapter 4

Importance of Plant Diseases

Introduction

Plant diseases have long been a significant concern in agriculture, horticulture, and ecosystems worldwide. These ailments affecting plant life, caused by various pathogens such as fungi, bacteria, viruses, and nematodes, have far-reaching consequences that extend beyond the fields and gardens where they manifest.

The importance of plant diseases cannot be overstated, as they impact global food security, economic stability, and the delicate balance of our natural environments. In this regard, understanding the profound significance of plant diseases is not only a matter of agricultural or botanical concern but a crucial aspect of sustaining our planet's well-being and ensuring the nourishment of its inhabitants. This introduction sets the stage for a deeper exploration of the multifaceted importance of plant diseases in the intricate web of life.

Plant diseases are of significant importance in agriculture, ecology, and the global food supply. They can have far-reaching consequences on both natural ecosystems and human livelihoods.

Understanding the Importance of Plant Diseases

Economic Impact

Plant diseases pose a substantial threat to global agriculture and food security. They can result in reduced crop yields, lower crop quality, and, in severe cases, complete crop failure. This directly affects the livelihoods of farmers and can lead to food shortages and increased food prices, impacting both rural and urban populations. The economic losses due to plant diseases are enormous, with billions of dollars being spent annually on pest and disease management, including the development of resistant crop varieties and the application of pesticides.

Ecological Consequences

Plant diseases can disrupt ecosystems and have a cascading effect on the environment. When plant populations are weakened or killed by diseases, it can impact herbivores and other organisms that depend on these plants for food and habitat. In some cases, the loss of plant species due to diseases can result in changes in the composition and structure of ecosystems, leading to shifts in biodiversity. Furthermore, the use of chemical pesticides to combat plant diseases can have negative consequences on non-target species, such as pollinators and beneficial insects, which further disrupts ecological balance.

Global Food Security

With a rapidly growing global population, ensuring food security is a top priority. Plant diseases threaten food security by reducing crop productivity. When diseases affect staple crops like wheat, rice, and maize, it can have profound consequences on human nutrition and well-being. To mitigate this threat, researchers and agricultural experts are continuously working on developing disease-resistant crop varieties and sustainable disease management strategies. The prevention and control of plant diseases are essential to ensure that enough food is produced to feed the world's population.

Plant diseases have a far-reaching impact, affecting not only the agricultural sector but also ecosystems and global food security. Recognizing their importance underscores the need for ongoing research, innovative solutions, and sustainable practices to manage and mitigate the risks associated with plant diseases.

Plant diseases hold a paramount importance in our world for several reasons, spanning the realms of agriculture, environment, and human society:

Agricultural Impact

Plant diseases pose a significant threat to global food security. Crop plants that succumb to diseases can lead to reduced yields, lower crop quality, and, in severe cases, crop failure. This directly affects the livelihoods of farmers, the availability and affordability of food for consumers, and can even result in economic losses on a national or global scale.

Economic Consequences

Beyond agriculture, plant diseases have substantial economic implications. They result in increased production costs for farmers due to the need for disease management strategies, such as pesticides and fungicides. These costs can be transferred to consumers through higher prices, affecting household budgets and overall economic stability.

Biodiversity

Plant diseases can influence ecosystems and the natural world by affecting native plant species and the animals that depend on them. Invasive pathogens can devastate entire ecosystems by wiping out dominant plant species, which can have cascading effects on the animals and insects that rely on those plants for food and shelter.

Environmental Balance

Plants play a crucial role in maintaining environmental balance. They help in carbon sequestration, oxygen production, and water filtration. Plant diseases can disrupt these essential functions, impacting air and water quality, and exacerbating issues such as climate change.

Human Health

Some plant diseases can indirectly impact human health by affecting the plants that are part of our diet. Consuming diseased crops may lead to the ingestion of toxins or pathogens, potentially causing health issues.

International Trade

The global trade of plants and plant products is vast, and plant diseases can be easily spread via these routes. This can lead to trade restrictions and barriers as countries aim to protect their agricultural industries from the introduction of harmful pathogens.

Research and Innovation

The study of plant diseases has led to significant advancements in genetics, biology, and technology. Understanding how pathogens affect plants has driven research in biotechnology, genetics, and crop breeding, resulting in the development of disease-resistant crop varieties.

Education and Awareness

Plant diseases also provide valuable educational opportunities. They underscore the importance of responsible agricultural practices, biodiversity conservation, and the need for proactive disease management strategies. This awareness is crucial for sustainable agriculture and ecosystem preservation.

The importance of plant diseases extends far beyond the confines of farms and gardens. They have far-reaching implications, affecting our food supply, economies, ecosystems, and the health of our planet. Understanding and effectively managing plant diseases are essential for the well-being of both human society and the environment.

Scope of Plant Pathology

- ☆ Conducting field surveys in fields, orchards, and other areas to determine the prevalence and extent of plant diseases.
- ☆ Identifying and documenting new economically significant diseases, including their causes and extent of occurrence.
- ☆ Evaluating the economic losses caused by various plant diseases of importance.
- ☆ Investigating the causes, symptoms, contributing factors, and recurrence patterns of these diseases.
- ☆ Developing cost-effective methods for managing economically significant plant diseases.
- ☆ Contributing to the development of disease-resistant plant varieties through research and collaboration.
- ☆ Providing training to extension workers and subject matter specialists to facilitate effective communication between pathologists and farmers, promoting improved crop production practices.

Chapter 5

Classification of Plant Diseases

Introduction

Plant diseases are a significant concern in agriculture, horticulture, and forestry, as they can lead to devastating consequences such as reduced crop yields, economic losses, and food security issues. Understanding and classifying plant diseases is a fundamental step in their management and control.

The classification of plant diseases is a systematic approach that categorizes them based on various criteria, providing valuable insights into their causes, symptoms, and potential control strategies. This comprehensive classification system helps researchers, plant pathologists, and farmers to identify and manage diseases effectively, making it an essential tool in the realm of plant health and agriculture.

Plant diseases can be classified into several categories based on the following criteria:

- ☆ **Causal Agent:** Plant diseases can be caused by a variety of pathogens, and one of the primary classifications is based on the causal agent. This includes fungal diseases, bacterial diseases, viral diseases, parasitic diseases, and nematode diseases.
- ☆ **Affected Plant Part:** Diseases may target different parts of the plant, leading to the classification of foliar diseases, stem and branch diseases, root diseases, and fruit and flower diseases.
- ☆ **Symptoms:** The symptoms expressed by diseased plants can vary widely, with some diseases causing necrosis (tissue death), others

inducing chlorosis (yellowing), and some leading to hypertrophy (abnormal growth). This classification aids in the identification of diseases by their visual characteristics.

- **Mode of Transmission:** Understanding how diseases spread is crucial for disease management. Diseases can be classified as soilborne, airborne, or vector-borne, depending on their mode of transmission.
- **Host Range:** Some diseases are highly specific to certain plant types, while others affect a broad range of hosts. This classification helps in predicting the potential impact of a disease in a given area.
- **Environmental Conditions:** Some diseases are favored by specific environmental conditions, while others thrive under stress-induced scenarios, such as drought or nutrient deficiencies.
- **Control Methods:** Classifying diseases based on the available control methods, such as chemical, biological, cultural, or genetic control, provides insights into the most effective strategies for disease management.
- **Emerging and Reemerging Diseases:** The classification also accounts for diseases that may appear in new areas or make a comeback due to changing environmental conditions, host susceptibility, or other factors. Surveillance and monitoring are vital in addressing such diseases.

A systematic classification of plant diseases not only aids in their diagnosis and management but also supports research and policy decisions in agriculture. By understanding the diverse nature of plant diseases and their causes, plant pathologists and agriculturists can develop more effective disease control strategies, ultimately safeguarding global food production and agricultural sustainability.

Classification of Plant Diseases

Plant diseases pose significant threats to agricultural productivity and ecosystem stability, impacting global food security and biodiversity. The classification of plant diseases is a crucial framework that aids researchers, agronomists, and plant pathologists in understanding, managing, and preventing these ailments. The intricate relationships between plants and pathogens have led to a diverse range of diseases caused by fungi, bacteria, viruses, nematodes, and other agents. By categorizing plant diseases based on their causal agents, symptoms, and modes of transmission, scientists can develop targeted strategies for disease control and mitigation. This classification system serves as a fundamental tool for studying the dynamic interactions within plant ecosystems, enabling the development of sustainable agricultural practices and the preservation of vital plant species.

Classification of Plant Diseases by Causal Agent

Plant diseases are often classified by their physiological effects or symptoms. Many diseases, however, produce practically identical symptoms and signs but are caused by very different microorganisms or agents, thus requiring completely different control methods. Classification according to symptoms is also inadequate because a causal agent may induce several different symptoms, even on the same plant organ, which often intergrade. Classification may be according to the species of plant affected. Host indexes (lists of diseases known to occur on certain hosts in regions, countries, or continents) are valuable in diagnosis. When an apparently new disease is found on a known host, a check into the index for the specific host often leads to identification of the causal agent. It is also possible to classify diseases according to the essential process or function that is adversely affected. The best and most widely used classification of plant diseases is based on the causal agent, such as a noninfectious agent or an infectious agent (*i.e.*, a virus, viroid, mycoplasma, bacterium, fungus, nematode, or parasitic flowering plant).

Noninfectious Disease-Causing Agents

Noninfectious diseases in plants are caused by various agents that differ from infectious pathogens such as bacteria, viruses, fungi, and nematodes. These noninfectious agents can lead to diseases through environmental factors, physiological imbalances, or genetic abnormalities. Noninfectious diseases, which sometimes arise very suddenly, are caused by the excess, deficiency, nonavailability, or improper balance of light, air circulation, relative humidity, water, or essential soil elements; unfavourable soil moisture-oxygen relations; extremes in soil acidity or alkalinity; high or low temperatures; pesticide injury; other poisonous chemicals in air or soil; changes in soil grade; girdling of roots; mechanical and electrical agents; and soil compaction. In addition, unfavourable preharvest and storage conditions for fruits, vegetables, and nursery stock often result in losses. The effects of noninfectious diseases can be seen on a variety of plant species growing in a given locality or environment. Many diseases and injuries caused by noninfectious agents result in heavy loss but are difficult to check or eliminate because they frequently reflect ecological factors beyond human control. Symptoms may appear several weeks or months after an environmental disturbance. Injuries incurred from accidents, poisons, or adverse environmental disturbances often result in damaged tissues that weaken a plant, enabling bacteria, fungi, or viruses to enter and add further damage. The cause may be obvious (lightning or hail), but often it is obscure. Symptoms alone are often unreliable in identifying the causal factor. A thorough examination of recent weather patterns, the condition of surrounding plants, cultural treatments or disturbances, and soil and water tests can help reveal the nature of the disease.

Adverse Environment

High temperatures may scald corn, cotton, and bean leaves and may induce formation of cankers at the soil surface of tender flax, cotton, and peanut plants. Frost injury is relatively common, but temperatures just above freezing also may cause damage, such as net necrosis (localized tissue death) in potato tubers and "silvering" of corn leaves. Isolated, thin-barked trees growing in northern climates and subjected to frequent thawing by day and freezing by night may develop dead bark cankers or vertical frost cracks on the south or southwest sides of the trunk. Alternate freezing and thawing, heaving, low air moisture, and smothering under an ice-sheet cover are damaging to alfalfa, clovers, strawberries, and grass on golf greens. Legume crowns commonly split under these conditions and are invaded by decay-forming fungi. The drought and dry winds that often accompany high temperatures cause stunting, wilting, blasting, marginal scorching of leaves, and dieback of shoots. Leaf scorch is common on trees in exposed locations following hot, dry, windy weather when water is lost from leaves faster than it is absorbed by roots. Leaf scorch and sudden flower drop are common indoor plant problems because the humidity in a home, an apartment, or an office is usually below 30 percent. Similar symptoms are caused by a change in soil grade, an altered water-table level, a compacted and shallow soil, paved surface over tree roots, temporary flooding or a waterlogged (oxygen-deficient) soil, girdling tree roots, salt spray near the ocean, and an injured or diseased root system. Injured plants are often very susceptible to air and soil pathogens and secondary invaders. Blossom-end rot of tomato and pepper is prevalent when soil moisture and temperature levels fluctuate widely and calcium is low.

Poor aeration may cause blackheart in stored potatoes. Accumulation of certain gases from the respiration of apples in storage may produce apple scald and other disorders. All plants require certain mineral elements to develop and mature in a healthy state. Macronutrients such as nitrogen, potassium, phosphorus, sulfur, calcium, and magnesium are required in substantial quantities, while micronutrients or trace elements such as boron, iron, manganese, copper, zinc, and molybdenum are needed in much smaller quantities. When the supply of any essential nutrient falls below the level required by the plant, a deficiency occurs, leading to symptoms that include stunting of plants; scorching or malformation of leaves; abnormal coloration; premature leaf, bud, and flower drop; delayed maturity or failure of flower and fruit buds to develop; and dieback of shoots.

Symptoms of nutrient deficiencies vary depending on the nutrients involved, the stage of plant growth, soil moisture, and other factors; they often resemble symptoms caused by infectious agents such as bacteria or viruses. The availability of water may affect nutrient uptake by the plant. Blossom-end rot of tomato, a disease associated with a deficiency of calcium, may occur if the water supply is irregular, even if an adequate amount of calcium is in the soil.

This discontinuity in availability of water will inhibit uptake of the calcium in a quantity sufficient to nourish a fast-growing tomato plant. Necrosis at the blossom end of the fruit results. This situation generally disappears when water conditions improve. Excess minerals can damage plants either directly, causing stunting, deformities, or dieback, or indirectly by interfering with the absorption and use of other nutrients, resulting in subsequent deficiency symptoms. A superabundance of nitrogen, for example, may cause deficiency symptoms of potassium, zinc, or other nutrient elements; a lack of or delay in flower and fruit development; and a predisposition to winter injury. If potassium is high, calcium and magnesium deficiencies may occur.

The pH of a soil has a dramatic impact on nutrient availability to plants. Most plants will grow in a soil with a pH between 4.0 and 8.0. In acidic soils some nutrients are far more available and may reach concentrations that are toxic or that inhibit absorption of other nutrients, while other minerals become chemically bound and unavailable to plants. A similar situation exists in alkaline soils, although different minerals are affected. Oats planted in alkaline soils that actually contain a sufficient amount of manganese may develop the manganese-deficiency disease gray speck. This occurs because an elevated soil pH causes manganese to react with oxygen to produce manganese dioxide, a form of the nutrient that is insoluble to plants.

An excess of water-soluble salts is a common problem with houseplants. Salt concentrations may build up as a whitish crust on soil and container surfaces of potted plants following normal evaporation of water over a period of time. Symptoms include leaf scorching, bronzing, yellowing and stunting, and wilting, plus root and shoot dieback. Damage from soluble salts is also common in arid regions and in regions where ice-control chemicals are applied heavily. Several nonparasitic diseases (*e.g.*, oat blast, weakneck of sorghum, straighthead of rice, and crazy-top of cotton) are caused by combinations of environmental factors–*e.g.*, high temperatures, moisture stress or poor irrigation practices, imbalance of mineral nutrients, and reduced light. Environmental disturbances alter the normal physiology of the plant, activity of pathogens, and host-pathogen interactions.

Toxic Chemicals

Many complex chemicals are routinely applied to plants to prevent attack by insects, mites, and pathogens; to kill weeds; or to control growth. Serious damage may result when fertilizers, herbicides, fumigants, growth regulators, antidesiccants, insecticides, miticides, fungicides, nematicides, and surfactants (substances with enhanced wetting, dispersing, or cleansing properties, such as detergents) are applied at excessive rates or under hot, cold, or slow-drying conditions. Some pollutants are the direct products of industry and fuel combustion, while others are the result of photochemical reactions between products of combustion and naturally occurring atmospheric compounds.

The major pollutants toxic to plants are sulfur dioxide, fluorine, ozone, and peroxyacetyl nitrate.

Sulfur dioxide results primarily from the burning of large amounts of soft coal and high-sulfur oil. It is toxic to a wide range of plants at concentrations as low as 0.25 part per million (ppm) of air (*i.e.*, on a volume basis, one part per million represents one volume of pure gaseous toxic substance mixed in one million volumes of air) for 8 to 24 hours. Gaseous and particulate fluorides are more toxic to sensitive plants than is sulfur dioxide because they are accumulated by leaves. They are also toxic to animals that feed on such foliage. Fluorine injury is common near metal-ore smelters, refineries, and industries making fertilizers, ceramics, aluminum, glass, and bricks. Ozone and peroxyacetyl nitrate injury (also called oxidant injury) are more prevalent in and near cities with heavy traffic problems. Exhaust gases from internal combustion engines contain large amounts of hydrocarbons (substances that principally contain carbon and hydrogen molecules–gasoline, for example). Smaller amounts of unconsumed hydrocarbons are formed by combustion of fossil fuels (*e.g.*, coal, oil, natural gas) and refuse burning. Ozone, peroxyacetyl nitrate, and other oxidizing chemicals (smog) are formed when sunlight reacts with nitrogen oxides and hydrocarbons. This pollutant complex is damaging to susceptible plants many kilometres from its source. Ozone and peroxyacetyl nitrate are capable of causing injury if present at levels of 0.01 to 0.05 part per million for several hours.

Physical Injury

Lightning, hail, high winds, ice and snow loads, machinery, insect and animal feeding, and various cultural practices may seriously injure plants or plant products. With the exception of lightning, which may cause death of trees and succulent crop plants in limited areas, such injury does not usually kill plants. Wounds are created, however, through which pathogens may enter.

Infectious Disease-Causing Agents

Plants are subject to infection by thousands of species from very diverse groups of organisms. Most are microscopic, but a few are macroscopic. The infectious agents, as previously mentioned, are called pathogens and can be grouped as follows: viruses and viroids, bacteria (including mycoplasmas and spiroplasmas, collectively referred to as mycoplasma-like organisms [MLOs]), fungi, nematodes, and parasitic seed plants.

Diseases Induced by Viruses and Viroids

Viruses and viroids are two distinct types of pathogens that can cause diseases in plants. While viruses are composed of genetic material (either DNA or RNA) surrounded by a protein coat, viroids consist solely of a short, single-stranded RNA molecule without a protective protein coat. Both of these pathogens can lead to a variety of diseases in plants, impacting agricultural productivity and ecosystem health.

Viruses and viroids are the smallest of the infectious agents. The structurally mature infectious particle is called a virion. Virions range in size from approximately 20 nanometres (0.0000008 inch) to 250–400 nanometres and are of various shapes. Viroids differ from viruses in that they have no structural proteins, such as those that form the protein coat (capsid) of the virus.

Both viruses and viroids are obligate parasites–*i.e.*, they are able to multiply or replicate only within a living cell of a particular host. A single plant species may be susceptible to infection by several different viruses or viroids. Major disease of important food crops such as potato, tomato, wheat, oats, rice, corn, peach, orange, sugar beet, sugarcane, and palm result from viral infection. Diseases are generally most serious in plants that are propagated vegetatively, or asexually–*i.e.*, grown from cuttings, cut divisions, sprouts, and other plant material–rather than grown from seeds (sexually propagated).

Symptoms

Tomato Leaves Puckered and Blistered by the Tobacco Mosaic Virus.

The symptoms of viral and viroid plant diseases fall into four groups: (1) change in colour–yellowing, green and yellow mottling, and vein clearing; (2) malformations–distortion of leaves and flowers, rosetting, proliferation and witches'-brooms (abnormal proliferation of shoots), and little or no leaf development between the veins; (3) necrosis–leaf spots, ring spots, streaks, wilting or drooping, and internal death, especially of phloem (food-conducting) tissue; and (4) stunting or dwarfing of leaves, stems, or entire plants. Rarely they may kill the host in a short time (*e.g.*, spotted wilt and curly top of tomato). More commonly they cause reduced yield and lower quality of product.

In many cases, virus-infected plants are more susceptible to root rots, stem or stalk rots, seedling blights, and possibly other types of diseases.

Some plants may carry one or more viruses and show no symptoms; thus, they are latent carriers and a source of infection for other plants. Symptoms of certain virus-infected plants, such as geraniums, may be masked at high temperatures. Virus symptoms reappear when the weather cools.

For convenience, viral/viroid diseases are often grouped together generally by symptoms, regardless of true viral/viroid relationships. Viruses also can be grouped into strains, each differing greatly in virulence and other properties. For example, two virus strains, chemically distinct, may produce indistinguishable symptoms in one orchid plant but strikingly different symptoms in another. Diseases caused by unrelated viruses may resemble one another more closely than diseases caused by strains of the same virus. Certain variegated plants, such as Abutilon and Rembrandt tulips, owe their horticultural uniqueness and desirability to being inherently virus-infected.

Transmission

With the exception of tobacco mosaic virus, relatively few viruses or viroids are spread extensively in the field by contact between diseased and healthy leaves. All viruses that spread within their host tissues (systemically) can be transmitted by grafting branches or buds from diseased plants on healthy plants. Natural grafting and transmission are possible by root grafts and with parasitic dodder (Cuscuta species). Vegetative propagation often spreads plant viruses. Fifty to 60 viruses are transmitted in seed, and a few seed-borne viruses, such as sour-cherry yellows, are carried in pollen and transmitted by insects.

Most disease-causing viruses are carried and transmitted naturally by insects and mites, which are called vectors of the virus. The principal virus-carrying insects are about 200 species of aphids, which transmit mostly mosaic viruses, and more than 100 species of leafhoppers, which carry yellows-type viruses. Whiteflies, thrips, mealybugs, plant hoppers, grasshoppers, scales, and a few beetles also serve as vectors for certain viruses. Some viruses may persist for weeks or months and even duplicate themselves in their insect vectors; others are carried for less than an hour. Slugs, snails, birds, rabbits, and dogs also transmit a few viruses, but this is not common.

A small number of plant viruses are soilborne. Viruses causing grape fanleaf, tobacco rattle, and tobacco and tomato ring spots, as well as several strawberry viruses, are spread by nematodes feeding externally (*i.e.*, ectoparasitic) on plant roots. A few soilborne viruses may be spread by the swimming spores of primitive, soil-inhabiting pathogenic fungi, such as those causing big vein of lettuce, soilborne wheat mosaic, and tobacco necrosis. Viruses often overwinter in biennial and perennial crops and weeds (plants that overwinter by means of roots and produce seed in their second year or during several years, respectively), in plant debris, and in insect vectors. Plants, once infected, normally remain so for life.

Control Measures

After a plant is infected with a virus/viroid, little can be done to restore its health. Control is accomplished by several methods, such as growing resistant species and varieties of plants or obtaining virus-free seed, cuttings, or plants as a result of indexing and certification programs. Indexing is a procedure to determine the presence or absence of viruses not readily transmitted mechanically. Material from a "test" plant is grafted to an "indicator" plant that develops characteristic symptoms if affected by the viral disease in question. In addition, more drastic measures are sometimes followed, including destroying (roguing) infected crop and weed host plants and enforcing state and national quarantines or embargoes. Further control measures include controlling insect vectors by spraying plants with contact insecticides or fumigating soil to kill insects, nematodes, and other possible vectors. Growing valuable plants under fine cheesecloth or wire screening that excludes insect vectors also is done. Separation of new from virus-infected plantings of the same or closely related species is sometimes effective, and the simple practice of not propagating from plants suspected or known to harbour a virus also reduces loss.

Infected peach, apple, and rose budwood stock and carnations have been grown for weeks or months at temperatures about 37 to 38 °C (99 to 100 °F) to free new growth from viruses. Soaking some woody plant parts or virus-infected sugarcane shoots in hot water at about 50 °C (120 °F) for short periods also is effective. Both dry and wet heat treatments are based on the sensitivity of certain viruses to high temperatures. Rapidly growing dahlia and chrysanthemum sprouts outgrow viruses so that stem tips can be used to propagate healthy plants. With certain carnations, chrysanthemums, and potatoes, a few cells from the growing tip have been grown under sterile conditions in tissue culture; from these, whole plants have been developed free from viruses. Examples of virus and viroid diseases are characterized in the Table 5.1.

Diseases Caused by Bacteria

Bacterial infections can give rise to a myriad of diseases, ranging from common ailments to more severe and potentially life-threatening conditions. These microscopic organisms, ubiquitous in our environment, possess the ability to invade various tissues and systems within the human body, triggering a diverse array of health issues. Understanding the nature of bacterial diseases is crucial for developing effective prevention strategies, diagnostics, and treatments to safeguard public health. In this exploration, we will delve into some of the

Table 5.1: Viral and Viroid Diseases of Plants

Disease	*Causative Agent*	*Hosts*	*Symptoms and Signs*	*Additional Features*
Tobacco mosaic	Tobacco mosaic virus (TMV)	Tobacco, tomato, and hundreds of other vegetables and weeds	Mottled appearance of leaves (mosaic pattern); dwarfing	Virus remains viable for years in soil and tobacco; the disease occurs worldwide; significant economic losses can occur
Cucumber mosaic	Cucumber mosaic virus (CMV)	Cucumber, bean, tobacco, and other plants (wide range of hosts)	Similar to those of TMV infections	Worldwide occurrence; very broad range of hosts
Barley yellow dwarf	Barley yellow dwarf virus (BYDV)	Barley, oats, rye, wheat; also pasture grasses and weeds	Yellowing and dwarfing of leaves; stunting of plants	One of the most important diseases of small grains
Tomato spotted wilt	Tomato spotted wilt virus (TSWV)	Tomato, pepper, pineapple, peanut, and many other plants	Leaves show concentric, necrotic rings; necrotic region yellow, then turning red-brown	Very wide host range; infects hundreds of different plants
Prunus necrotic ring spot	Prunus necrotic ring spot virus (PNRV)	Stone fruits—*e.g.*, cherry, almond, peach, apricot, plum, and others	Delayed foliation; leaves on infected branches show light green spots and dark rings, then become necrotic and fall off	Very widespread disease of stone fruits; affects almost all trees in fruit-producing regions
Potato spindle tuber	Potato spindle tuber viroid (PSTV)	Potato and tomato	Stunted growth; tubers are spindle-shaped and smaller than healthy tubers	The first identified viroid infection in plants; can cause major reduction in crop yield
Citrus exocortis	Citrus exocortis viroid (CEV)	Orange, lemon, lime, and other citrus plants	Infected trees show vertical splits in bark, thin strips of partially loosened bark, and a cracked, scaly appearance	Worldwide distribution; causes reduction of crop yield

notable diseases caused by bacteria, examining their symptoms, transmission methods, and the ongoing efforts to combat these microbial adversaries.

Thousands of bacterial species occur in nature. Many of these perform biochemical processes essential for the continuity of life; for example, bacterial detritivores, or decomposers, feed on nonliving organic matter, recycling it through the ecosystem. There are, however, hundreds of bacterial species that cause diseases in humans, animals, and plants.

Bacteria are prokaryotic microorganisms–*i.e.*, single-celled microorganisms in which the nuclear substance is not enclosed in a membrane. There are two major types of bacteria, the eubacteria and the archaebacteria, and they are distinguished by differences in the composition of their cell wall and cytoplasmic membrane and by certain metabolic features. Plant pathogens belong to the eubacteria. The eubacteria can be divided into three groups: gram-negative bacteria, gram-positive bacteria, and the mycoplasmas and spiroplasmas, referred to as mycoplasma-like organisms (MLOs). Gram-negative and gram-positive bacteria are distinguished on the basis of their cell wall structure, which affects the ability of the bacterium to react to the Gram stain–one of the most useful stains in bacteriologic laboratories. The distinguishing characteristic of MLOs is their lack of a cell wall; their outer boundary is instead a cytoplasmic membrane, which imparts some unusual properties not found in most eubacteria. MLOs belong to the taxonomic class Mollicutes. Plant diseases caused by MLOs are grouped as agents of "decline" (characterized by loss of vigour, decrease in yield of fruit, and eventual death) and agents of virescence (the greening of flowers) and developmental abnormalities.

The principal genera of plant pathogenic bacteria are *Agrobacterium, Clavibacter, Erwinia, Pseudomonas, Xanthomonas, Streptomyces*, and *Xylella*. With the exception of Streptomyces species, all are small, single, rod-shaped cells approximately 0.5 to 1.0 micrometre (0.00002 to 0.00004 inch) in width and 1.0 to 3.5 micrometres in length. Streptomycetes develop branched mycelia (narrow, threadlike growth) with curled chains of conidia (spores) on the tips of the mycelia. Streptomyces are gram-positive; most species of the other genera are gram-negative.

Symptoms and Signs

Bacterial diseases can be grouped into four broad categories based on the extent of damage to plant tissue and the symptoms that they cause, which may include vascular wilt, necrosis, soft rot, and tumours. Vascular wilt results from the bacterial invasion of the plant's vascular system. The subsequent multiplication and blockage prevents movement (translocation) of water and nutrients through the xylem of the host plant. Drooping, wilting, or death of the aerial plant structure may occur; examples include bacterial wilt of sweet corn, alfalfa, tobacco, tomato, and cucurbits (*e.g.*, squash, pumpkin, and cucumber) and black rot of crucifers. Pathogens can cause necrosis by secreting a toxin (poison).

Symptoms include formation of leaf spots, stem blights, or cankers. Soft rot diseases are caused by pathogens that secrete enzymes capable of decomposing cell wall structures, thereby destroying the texture of plant tissue–*i.e.*, the plant tissue becomes macerated (soft and watery). Soft rots commonly occur on fleshy vegetables such as potato, carrot, eggplant, squash, and tomato. Tumour diseases are caused by bacteria that stimulate uncontrolled multiplication of plant cells, resulting in the formation of abnormally large structures.

Most bacteria produce one major symptom, but a few produce a range or combination of symptoms. In general, it is not particularly difficult to tell whether a plant is affected by a bacterial pathogen; however, identification of the causative agent at the species level requires isolation and characterization of the pathogen using numerous laboratory techniques.

Transmission and Infection

In order for a bacterium to produce a disease in a plant, the bacterium must first invade the plant tissue and multiply. Bacterial pathogens enter plants through wounds, principally produced by adverse weather conditions, humans, tools and machinery, insects, and nematodes, or through natural openings such as stomata, lenticels, hydathodes, nectar-producing glands, and leaf scars.

Most foliage invaders are spread from plant to plant by windblown rain or dust. Humans disseminate bacteria through cultivation, grafting, pruning, and transporting diseased plant material. Animals, including insects and mites, are other common transmission agents. Some bacteria, such as the causal agent of Stewart's, or bacterial, wilt of corn (*Erwinia stewartii*), not only are spread by a flea beetle but also survive over winter in this insect.

When conditions are unfavourable for growth and multiplication, bacteria remain dormant on or inside plant tissue. Some, such as the crown gall bacterium, may survive for months or years in the soil.

Bacterial diseases are influenced greatly by temperature and moisture. Often, a difference of only a few degrees in temperature determines whether a bacterial disease will develop. In most cases, moisture as a water film on plant surfaces is essential for establishing an infection.

Control Measures

In general, the diseases caused by bacteria are relatively difficult to control. This is partly attributable to the speed of invasion as bacteria enter natural openings or wounds directly. Direct introduction also enables them to escape the toxic effects of chemical protectants. Losses from bacterial diseases are reduced by the use of pathogen-free seed grown in arid regions. Examples of diseases controlled by this method include bacterial blights of beans and peas, black rot of crucifers, and bacterial spot and canker of tomato. Seed treatment with hot water at about 50 °C (120 °F) is also effective for crucifers, cucurbits, carrot, eggplant, pepper, and tomato. Bactericidal seed compounds control some bacterial diseases, such as angular leaf spot of cotton, gladiolus scab, and soft rot of ornamentals. Rotation with nonhost crops reduces losses caused by wilt of alfalfa, blights of beans and peas, black rot of crucifers, crown gall, and bacterial spot and canker of tomato. Eradication and exclusion of host plants has been useful against citrus canker, angular leaf spot of cotton, fire blight, and crown gall. Resistant varieties of crop plants have been developed to reduce losses from wilts of alfalfa, corn, and tobacco; angular leaf spot of cotton and tobacco; and bacterial pustule of soybeans, among others. Protective insecticidal sprays help control bacterial diseases, such as wilts of sweet corn and cucurbits and soft rot of iris. Protective bactericidal sprays, paints, or drenches containing copper or antibiotics are used against bacterial blights of beans and celery, fire blight, crown gall, blackleg of delphinium, and hazelnut and walnut blights. Finally, sanitary measures–*i.e.*, clean plow down of crop refuse, destruction of volunteer plants and weeds, sterilization of pruning and grafting tools–as well as refraining from cultivating when foliage is wet, overhead watering and spraying of indoor plants, and late cutting or grazing of alfalfa and other crops, are useful in reducing the incidence of bacterial diseases. The characteristics of several plant diseases caused by bacteria are summarized in the Table 5.2.

Diseases Caused by Fungi

Fungal infections in plants result in a multitude of diseases that can adversely affect crop yields, agricultural ecosystems, and natural plant populations. These diseases, caused by various pathogenic fungi, encompass a wide range of symptoms, from leaf spots and wilting to more severe conditions like rot and blight. Understanding the dynamics of plant-fungal interactions is crucial for devising effective strategies to manage and mitigate the impact of these diseases on global food production and biodiversity. In this exploration, we will delve into some prominent fungal diseases affecting plants, examining their modes of infection, consequences for plant health, and ongoing efforts to develop sustainable solutions for disease control in agriculture and ecosystems.

Fungi cause the great majority, an estimated two-thirds, of infectious plant diseases. They include all white and true rusts, smuts, needle casts, leaf curls, mildew, sooty molds, and anthracnoses; most leaf, fruit, and flower spots; cankers; blights; scabs, root, stem, fruit, and wood rots; wilts; leaf, shoot, and bud galls; and many others. All economically important plants apparently are attacked by one or more fungi; often many different fungi may cause disease in one plant species.

The fungi represent an extremely large and diverse group of eukaryotic microorganisms. The cells, which contain a membrane-bound nucleus, are devoid of chlorophyll and have rigid cell walls. Fungi often have a plantlike vegetative body consisting of microscopic branching threadlike filaments of various lengths, called hyphae (singular hypha), some of which extend into the air while others penetrate the substrate on which they grow. The hyphae are

Table 5.2: Bacterial Diseases of Plants

Disease	*Causative Agent*	*Hosts*	*Symptoms and Signs*	*Additional Features*
Granville wilt	*Pseudomonas solanacearum*	Tobacco, tomato, potato, eggplant, pepper, and other plants	Stunting, yellowing, and wilting of parts above ground; roots decay and become black or brown	Occurs in most countries in temperate and semitropical zones; causes crop losses of hundreds of millions of dollars
Fire blight	*Erwinia amylovora*	Apple and pear	Blossoms appear water-soaked and shrivel; spreads to leaves and stems, causing rapid dieback	First plant disease proved to be caused by a bacterium
Wildfire of tobacco	*Pseudomonas syringae*	Tobacco	Yellowish green spots on leaves	Wildfire of tobacco occurs worldwide; causes losses in seedlings and field plants
Blight of beans	*Xanthomonas campestris*	Beans (common blight)	Yellowish green spots on leaves	Most phytopathogenic xanthomonads and pseudomonads cause necrotic spots on green parts of susceptible hosts; may be localized or systemic
	Pseudomonas syringae	Beans (brown spot)	Small water-soaked spots on lower side of leaves enlarge, coalesce, and become necrotic	
Soft rot	*Erwinia carotovora*	Many fleshy-tissue fruits—*e.g.*, cabbage, carrot, celery, onion	Soft decay of fleshy tissues that become mushy and soft	Occurs worldwide; causes major economic losses
Crown gall	*Agrobacterium tumefaciens*	More than 100 genera of woody and herbaceous plants	Initially a small enlargement of stems or roots usually at or near the soil line, increasing in size, becoming wrinkled, and turning brown to black	The conversion of a normal cell to one that produces excessive cell multiplication is caused by a plasmid (a small circular piece of DNA) carried by the pathogenic bacterium
Aster yellows	Mycoplasma-like organism (MLO)	Many vegetables, ornamentals, and weeds	Chlorosis; dwarfing malformations	Greatest losses suffered by carrots; transmission by leafhoppers
Citrus stubborn disease	*Spiroplasma citri* (MLO)	Citrus and stone fruits and vegetables	Chlorosis, yellowing of leaves, shortened internodes, wilting	First MLO pathogen of plant disease cultured

arranged into a network called a mycelium. It is the mass of the mycelium that gives fungal growth its characteristic "cottony" or "fuzzy" appearance. Fungi reproduce by a variety of methods, both asexual and sexual. They produce many kinds of spores in very large numbers. For example, the colour of a moldy piece of bread is due to the colour of a massive number of microscopic mold spores.

Symptoms and Signs

In general, a fungal infection can cause local or extensive necrosis. It can also inhibit normal growth (hypotrophy) or induce excessive abnormal growth (hypertrophy or hyperplasia) in a portion of or throughout an entire plant.

Symptoms associated with necrosis include leaf spots, blight, scab, rots, damping-off, anthracnose, dieback, and canker. Symptoms associated with hyperplasia include clubroot, galls, warts, and leaf curls.

In some instances, the fungus infecting the plant may produce growth or structures on the plant, stems, or leaves such as masses of mycelium or aggregates of spores with a characteristic appearance. These developments are referred to as signs of infection, in contrast to symptoms, which refer specifically to the plant or plant tissue.

Transmission

Fungi are spread primarily by spores, which are produced in abundance. The spores can be carried and disseminated by wind currents, water (splashing and rain), soil (dust), insects, birds, and the remains of plants that once were infected. Vegetative fungal cells that exist in dead plant material also can be transmitted when they come in contact with a susceptible host.

The survival of vegetative cells of plant pathogenic fungi in nature depends on climatic conditions, particularly temperature and moisture. Vegetative cells can survive temperatures from –5 to 45 °C (23 to 113 °F); fungal spores are considerably more resistant. The germination of spores, however, is favoured by mild temperatures and high humidity.

Control Measures

Because many thousands of fungal species can infect a broad range of plants and because each fungal species has different characteristics, a variety of practices are available to control fungal diseases.

The principal control measures include the use of disease-free seed and propagating stock, the destruction of all plant materials that may harbour pathogenic fungi, crop rotation, the development and use of resistant plant varieties, and the use of chemical and biological fungicides.

Several fungal diseases are characterized in the Table 5.3.

Table 5.3: Fungal Diseases of Plants

Disease	*Causative Agent*	*Hosts*	*Symptoms and Signs*	*Additional Features*
Late blight of potato	*Phytophthora infestans*	Potato	Water-soaked dark green to black or purplish lesions with pale green margins on lower leaves, white mildew at edge of lesions	Responsible for Irish famine; caused starvation and death and mass migration of population
Chestnut blight	*Endothia parasitica*	Chestnut tree	Yellowish to reddish brown patches appear on bark; lesions spread quickly and girdle twigs or limbs, which die	Disease accidentally imported from Asia; first observed in New York in 1904 and rapidly spread across the United States, practically eliminating native American chestnuts
Dutch elm disease	*Ceratocystis ulmi*	Elm tree	Leaves wilt, turn dull green to yellow or brown, and drop off; branches die	The causative fungus is believed to have entered Europe from Asia during World War I and was later transported to the United States (1930) on elm burl logs imported for furniture veneer; elm bark beetles spread the pathogen in the United States
Black stem rust of wheat	*Puccinia graminis*	Wheat; many grasses	On wheat, rust-coloured pustules with spores, chlorosis of surrounding tissue, followed by development of black teliospores; on barberry, chlorosis and hypertrophy of infected tissue, orange spore masses	Disease occurs wherever wheat is grown; in 1935 it destroyed about 60 percent of the total hard red spring wheat crop in Minnesota and South Dakota; fungus has a complex life cycle, partly on wheat and partly on the barberry plant; eradication of the barberry plant is an important control measure
Coffee rust	*Hemileia vastatrix*	Coffee	Orange-yellow powdery spots on lower side of leaves; centres turn brown and leaves fall	Most destructive disease of coffee; has caused devastating losses in all coffee-producing countries

Disease	*Causative Agent*	*Hosts*	*Symptoms and Signs*	*Additional Features*
White-pine blister rust	*Cronartium ribicola*	White pine tree	Small, discoloured, spindle-shaped cankers surrounded by narrow band of yellow-orange bark; blisters exude secretion followed by bright orange pustules	One of the most important forest diseases in the United States; currant is the alternate host, and its eradication is an important control measure
Corn smut	*Ustilago maydis*	Corn	Minute galls form on young corn seedlings; on older plants, large galls are produced on the silk of ears and on tassels, leaves, and stalks	Occurs wherever corn is grown; may cause serious crop damage
Loose smut	*Ustilago nuda*	Barley, oats, wheat	Infected heads are covered with masses of olive-green spores	Worldwide occurrence; destroys kernels of the infected plant
Downy mildew	Many species of the family Peronosporaceae	Many types of plants: grapes, grasses, vegetables, and others	Yellow irregular spots appear on upper leaf surface; downy fungus growth appears on underside; leaves die	One of the first plant diseases controlled by a fungicide—*i.e.*, Bordeaux mixture, a mixture of lime and copper sulfate used on grapes
Powdery mildew	Many species of the family Erysiphaceae	Many types of plants: grasses, vegetables, shrubs, and trees	Spots of powdery mildew growth that enlarge to cover leaves or other plant organs	One of the most common and widely spread plant diseases
Apple scab	*Venturia inaequalis*	Apple	Small olive-coloured areas appear on young leaves, later turn black, and may coalesce; black circular spots appear on fruit	Occurs almost everywhere apples are grown; infection reduces fruit size and quality
Black spot of rose	*Diplocarpon rosae*	Rose	Large circular black lesions on leaves; leaves turn yellow and fall off	Classified as an anthracnose, which affects leaves, stems, and fruits of many plants
Anthracnose of grape	*Elsinae ampelina*	Grape	(As above)	(As above)

Disease	*Causative Agent*	*Hosts*	*Symptoms and Signs*	*Additional Features*
Nectria canker	*Nectria galligena*	Apple and pear and many hardwood forest trees	Initially small circular brown areas that enlarge and become depressed with raised edges; callus tissue produced around canker	One of the most important diseases of pear, apple, and hardwood forest trees
Black knot of plum and cherry	*Plowrightia morbosum*	Plum and cherry	Small black knotty swellings on twigs and branches	Occurs primarily in the eastern half of the United States and New Zealand
Brown rot	*Monilinia fructicola*	Stone fruits	Brown spots on blossoms; twigs develop small sunken brown cankers; fruit develops brown spots that spread rapidly	Worldwide occurrence; can cause heavy losses both in orchards and in shipment
Soft rot	*Rhizopus* species	Flowers, fruits, and vegetables with fleshy organs	Tissues become soft with water-soaked appearance that often spreads rapidly, followed by development of fuzzy gray mycelium and black spores	Infection develops most rapidly on ripe fruits with favourable conditions (moderate temperature and high humidity)
Fusarium wilt of tomato	*Fusarium oxysporum*	Tomatoes	Leaves are bent down, growth is stunted, plant dies; dark streaks appear in vascular tissue	One of the most destructive diseases of tomato; entire fields can be destroyed
Wilts of vegetables, flowers, and some trees	*Verticillium* species	Cotton, potato, tomato, alfalfa, shade trees, and others	Similar to fusarium wilts; develops primarily in seedlings that die shortly after infection; older plants also are attacked	Worldwide distribution; the fungus infects hundreds of species of plants

Diseases Caused by Nematodes

Nematodes parasitic on plants are active, slender, unsegmented roundworms (also called nemas or eelworms). The great majority cannot be seen with the unaided eye, because they are very small and translucent. Practically all adult forms fall within the range of 0.25 to 2 millimetres in length. About 1,200 species cause disease in plants. Probably every form of plant life is fed upon by at least one species of nematode. They usually live in soil and attack small roots, but some species inhabit and feed in bulbs, buds, stems, leaves, or flowers.

Mode of Nematode Attack

Nematodes parasitic on plants obtain food by sucking juices from them. Feeding is accomplished through a hollow, needlelike mouthpart called a spear or stylet. The nematode pushes the stylet into plant cells and injects a liquid containing enzymes, which digest plant cell contents. The liquefied contents are then sucked back into the nematode's digestive tract through the stylet. Nematode feeding lowers natural resistance, reduces vigour and yield of plants, and affords easy entrance for wilt-producing or root rot-producing fungi or bacteria and other nematodes. Nematode-infested plants are weak and often appear to suffer from drought, excessive soil moisture, sunburn or frost, a mineral deficiency or imbalance, insect injury to roots or stems, or disease.

Common symptoms of nematode injury include stunting, loss of green colour and yellowing; dieback of twigs and shoots; slow general decline; wilting on hot, bright days; and lack of response to water and fertilizer. Feeder root systems are reduced; they may be stubby or excessively branched, often discoloured, and decayed. Winterkill of orchard trees, raspberries, strawberries, ornamentals, and other perennials is commonly associated with nematode infestations.

Root injury develops partly from the nematodes feeding on cells and partly from toxic salivary excretions of the parasite. Tissues often respond by producing either an enlargement or degeneration of cells; sometimes both occur.

Many nematodes are native and attack cultivated plants when their natural hosts are removed. Others have been introduced with seedling plants, bulbs, tubers, and particularly in soil balled around roots of infested nursery stock.

Nematodes may live part of the time free in soil around roots or in fallow gardens and fields. They tunnel inside plant tissues (endoparasites) or feed externally from the surface (ectoparasites) and may enter a plant through wounds or natural openings or by penetrating roots. All nematodes parasitic on plants require living plant tissues for reproduction. Nematodes are attracted to host roots by sensing either the heat given off by roots or the chemicals secreted by roots.

Most species require 20 to 60 days to complete a generation from egg through four larval stages to adult and back to egg. Some nematodes have only one generation a year but still produce several hundred offspring.

Soil populations and developmental rate of nematodes are affected by the length of the growing season; temperature; availability of water and nutrients; and moisture, type, texture, and structure of soil. Also important are populations of nematode-parasitic bacteria, viruses, some 50 different nematode-trapping fungi, protozoans, mites, flatworms, or other pests, and other nematodes. Toxic chemicals added to the soil or those secreted by plant roots; crop rotations and past cropping history; species, variety, age and nutrition of growing plants; and other factors are additional conditions that affect nematode populations.

Certain species live strictly in light, sandy soils; some build up high populations in muck soils; and a few seem to thrive in heavy soils. High populations and greater crop damage are much more common in light sandy soils than in heavy clay soils.

Many plant-infecting nematodes become inactive at temperatures between 5 and 15 °C (41 and 59 °F) and 30 and 40 °C (86 and 104 °F). The optimum for most is 20 to 30 °C (68 to 86 °F), but this varies greatly with the species, stage of development, activity, growth of the host, and other factors.

Nematodes may be found in plant tissues in large numbers. Hundreds of thousands may be present in infested roots or bulbs.

Cysts or Tiny Nodes on Soybean Plant Roots, Containing Eggs of Nematodes.

After a plant-infecting nematode has been accidentally introduced into a garden or field, several years pass before the population builds up sufficiently (*i.e.*, up to several billion or more active nematodes per hectare) to cause conspicuous symptoms in a large number of plants. This is because nematodes move very slowly through soil–rarely more than 75 cm (29.5 inches) a year. Nematodes are easily spread, however, by moving infested soil, plant parts, or contaminated objects–*e.g.*, tools and machinery, bags and other containers, running water, wind, clothing, shoes, animals, birds, and infested planting stock.

Root-knot nematodes (*Meloidogyne* species) are well known because of the conspicuous "knots," or gall-like swellings, they induce on roots. More than 2,000 kinds of higher plants are subject to their attack. Losses are often heavy, especially in warm regions with long growing seasons. Certain species, however, such as the northern root-knot nematode (*M.* hapla), are found where soil may freeze to depths of nearly a metre. Vegetables, cotton, strawberry, and orchard trees are commonly attacked. Garden plants and ornamentals frequently become infested through nursery stock.

Root-lesion nematodes (*Pratylenchus* species), cosmopolitan in distribution, are endoparasites that cause severe losses to hundreds of different crop and ornamental plants by penetrating roots and making their way through the tissues, breaking down the cells as they feed. They deposit eggs from which new colonies develop. After a root begins to decline in vigour, nematodes move into the soil in search of healthy roots. Lesions form in the root as fungi and bacteria enter damaged tissues, and root rot often occurs. Annual crops may succumb early in the season, but perennials and orchard trees may not decline for several years.

The golden nematode of potatoes (*Heterodera rostochiensis*) is a menace of the European potato industry. Great efforts have been made to control it. The speck-sized golden cysts that dot infested plant roots are the remains of female bodies. Each cyst may contain up to 500 eggs, which hatch in the soil over a period of up to 17 years. A chemical given off by potato and tomato roots stimulates hatching of the eggs.

A related, cyst-forming species, the sugar beet nematode (*H. schachtii*), is a pest that has restricted acreage of sugar beets in Europe, Asia, and America.

The citrus nematode (*Tylenchulus semipenetrans*) occurs wherever citrus is grown, exacting a heavy toll in fruit quality and production. Typical symptoms are a slow decline, yellowing and dying of leaves, and dieback of twigs and branches in many groves 15 years or older. Infested nursery stock has widely distributed the nematode. The burrowing nematode (*Radopholus similis*) is a serious endoparasite in tropical and subtropical areas, where it attacks citrus (causing spreading decline), banana, avocado, tomato, black pepper, abaca, and more than 200 important crops, trees, and ornamentals, causing severe losses.

Many important ectoparasites feed on plant roots–dagger nematodes (Xiphinema), stubby-root nematodes (Trichodorus), spiral nematodes (Rotylenchus and Helicotylenchus), sting nematodes (Belonolaimus), and pin nematodes (Paratylenchus). Leaf, or foliar, nematodes (*Aphelenchoides* species) and bulb and stem nematodes (*Ditylenchus dipsaci*) cause severe losses in vegetable and ornamental bulb crops, clovers, alfalfa, strawberry, sweet potato, orchids, chrysanthemums, begonias, and ferns.

Control Measures

Control measures for nematodes often include rotation with nonhost plants, growing of resistant varieties and species, use of certified, nematode-free nursery stock, and use of soil fumigants (nematicides) as preplanting or postplanting treatments.

Steam or dry heat is applied to soil in confined areas, such as greenhouse benches and ground beds. Exposure to moist heat, such as steam or hot water at 50 °C (120 °F) for 30 minutes, is sufficient to kill most nematodes and nematode eggs. Shorter periods are needed at higher temperatures. State and federal quarantines prohibiting movement of infested soil, plants or plant parts, machinery, and other likely carriers also exist.

Cultural practices to promote vigorous plant growth (*i.e.*, watering during droughts, proper application of fertilizers, clean cultivation, fall and summer fallowing, use of heavy organic mulches or cover crops, and plowing out roots of susceptible plants after harvest) are useful for specific nematodes. Asparagus, marigolds (*Tagetes* species), and *Crotalaria* species are toxic to many plant-infecting nematodes.

Parasitic Seed Plants

A number of flowering plants are parasites of other plants. Among the more important ones are mistletoe, dodder, and witchweed. Parasitic seed plants represent a fascinating and unique group of botanical organisms that have evolved distinct adaptations for their survival. Unlike conventional plants, these species rely on a parasitic lifestyle, tapping into the resources of host plants to sustain themselves. One notable example is the broomrape family (Orobanchaceae), which includes various species exhibiting parasitic tendencies. Broomrapes lack chlorophyll, rendering them unable to photosynthesize and produce their own nutrients. Instead, they utilize specialized structures called haustoria to attach to the roots of host plants, extracting water, minerals, and organic compounds. This parasitic relationship can have significant ecological implications, influencing plant community dynamics and challenging traditional notions of plant independence. The study of parasitic seed plants not only sheds light on their unique adaptations but also offers insights into the intricate interdependencies within plant ecosystems.

Mistletoe

Mistletoes are semiparasitic seed plants that feed on trees and obtain water and mineral salts by sending rootlike structures (haustoria) into vascular tissue of the inner bark. There are three important types: American (*Phorodendron* species), European (*Viscum album*), and dwarf (*Arceuthobium* species). All produce sticky seeds spread by birds. American mistletoe, restricted to the Americas, is best known for its ornamental and sentimental uses at Christmastime. The leafy, bushy evergreen masses, up to one metre or more in diameter, appear on tree branches. They are most conspicuous after deciduous leaves have fallen. The European mistletoe is similar in habit and appearance to its American relative. Tree branches infected by mistletoes may become stunted or even die.

Oak Mistletoe.

Dwarf mistletoe is common on and very destructive to conifers in forests. Seedlings and young trees may be stunted, deformed, or killed. Conspicuous witches'-brooms form in the crown or spindle-shaped swellings (later cankers) in limbs and trunk. Canker and wood-rotting fungi often enter through mistletoe wounds. Dwarf mistletoes frequently escape detection because the scaly-leaved plants may be less than 2.5 cm (1 inch) long; they do range to 30 to 45 cm (12 to 18 inches), however. Dwarf mistletoes occur scattered along conifer limbs and small branches. After the mistletoe has grown internally for about a year, the branch may start to form a witches'-broom. Four to five years elapse before the yellow to brown to olive-green shoots form fruits. The sticky seeds are shot with explosive force from the fruit for horizontal distances ranging from 5 to more than 18 metres (16 to more than 59 feet); this is one of the most remarkable methods for seed discharge among plants. Once seeds adhere to a branch, they germinate on young bark and penetrate into the host tree's vascular system. Control for mistletoes in individual trees involves removal of infected branches a foot or more beyond any evidence of the parasite before the fruits ripen.

Dodder

More than 100 species of dodder (Cuscuta) are widely distributed and called such names as strangleweed, devil's-hair, pull down, hell-bind, love vine, and goldthread. The leafless, yellow-orange, threadlike stems twine around a number of field and garden host plants. By extending to nearby plants, it may draw them together and downward until a tangled yellowish orange patch is formed. The infested area is usually less than three metres across the first year; it spreads more rapidly in succeeding years.

Parasitic Dodder.

Dodder is widely distributed as a contaminant with field seed; hence the losses in clover, alfalfa, and flax fields. Dodder is controlled by planting certified, properly cleaned seed and by mowing patches of dodder in the field well before the seeds form. The dried patches are sprinkled with fuel oil and burned. Careful application of selective herbicides or a soil fumigant and sowing heavily infested areas with resistant plants (*e.g.*, garden beans, soybean, corn, cowpea, pea, grasses, or small grains) are also control methods.

Witch Weed

Witchweed, scientifically known as Striga, is a notorious genus of parasitic plants that poses a significant threat to agricultural productivity, particularly in sub-Saharan Africa. Among the various species within this genus, Striga *hermonthica* stands out as a major concern, infesting crops such as maize, sorghum, and millet. Witchweed germinates in response to chemical signals released by the roots of host plants, initiating a parasitic relationship that siphons off nutrients and water from the host. This parasitism leads to stunted growth, wilting, and a drastic reduction in crop yields, contributing to food insecurity in vulnerable regions. Efforts to combat witchweed include developing resistant crop varieties, implementing cultural practices, and exploring biocontrol methods, as the persistence of this parasitic plant continues to challenge agricultural sustainability and the livelihoods of farmers in affected areas.

A serious infestation may cause corn plants to be severely stunted, wilt, and turn yellow or brown, thus reducing the acre yield. Striga plants, which rarely exceed heights of 20 to 25 cm (8 to 10 inches), have small red, yellowish red, yellow, or white flowers. One plant may produce hundreds of thousands of tiny brown seeds that can remain alive in soil for years until stimulated to germinate by a secretion from a nearby host root.

Witchweed.

Witchweed robs the host of water and food, causing it to grow more slowly than normal and often to die before maturing. Control is difficult; useful measures include application of selective herbicides before seeds are produced; rotation with a resistant crop and keeping plantings free of weed grasses that may serve as hosts; and prevention of seed set by growing trap crops and then destroying them with herbicides.

Chapter 6

Pathogenesis: Process of Disease Development

Introduction to Pathogenesis

Pathogenesis refers to the process by which a disease develops and progresses within an organism. It involves the study of the origin, development, and progression of a disease, including the cellular and molecular events that lead to its manifestation. Understanding pathogenesis is crucial for developing effective strategies for prevention, diagnosis, and treatment of diseases.

Key Concepts in Pathogenesis

Initiation of Disease

Diseases can be caused by various factors, including infectious agents (bacteria, viruses, fungi, parasites), genetic mutations, environmental factors (toxins, radiation), and autoimmune reactions.

The initiation of disease often involves a triggering event that disrupts normal cellular or physiological processes.

Host-Pathogen Interactions

In infectious diseases, pathogens interact with the host organism. This interaction is dynamic and involves the pathogen's ability to invade, survive, and multiply within the host, as well as the host's defense mechanisms to eliminate the pathogen.

The outcome of these interactions can result in various disease states, ranging from asymptomatic infection to severe illness.

Inflammation and Immune Response

The host's immune system plays a central role in the pathogenesis of many diseases. In response to infection or injury, the immune system triggers inflammation as a protective response.

Chronic inflammation can contribute to the development of various diseases, including autoimmune disorders and certain cancers.

Genetic and Environmental Factors

Genetic predisposition can influence an individual's susceptibility to certain diseases. Some conditions result from the interplay of genetic and environmental factors.

Environmental factors, such as diet, lifestyle, and exposure to pollutants, can contribute to the development and progression of diseases.

Cellular and Molecular Mechanisms

Pathogenesis involves changes at the cellular and molecular levels. This includes alterations in cell structure and function, disruptions in signaling pathways, and genetic mutations that contribute to disease development.

Understanding these mechanisms is essential for developing targeted therapies and interventions.

Clinical Manifestations

Pathogenesis ultimately leads to the clinical manifestations of a disease. These can include symptoms, signs, and laboratory abnormalities that are used for diagnosis.

The progression of a disease may involve distinct stages, with varying symptoms and severity over time.

Treatment and Prevention

Knowledge of pathogenesis is crucial for developing effective treatments. Therapeutic strategies often target specific steps in the disease process to alleviate symptoms, slow progression, or eliminate the underlying cause.

Preventive measures may also be designed based on an understanding of the factors contributing to disease development.

The pathogenesis encompasses the intricate processes through which diseases originate, progress, and manifest within living organisms. Researchers and healthcare professionals work to unravel these complex pathways to develop strategies for disease management and prevention.

Diagnosing Plant Diseases

Diagnosing plant diseases can be a complex process that involves careful observation, knowledge of plant pathology, and sometimes laboratory analysis.

The plant disease diagnosis depends on the exact determination of the disease cause. Generally, there are two plant disease causes: the pathogens and/or environmental factors. The former leads to infectious diseases, while the latter leads to noninfectious diseases.

Infectious Diseases

There are wide range of phytopathogens which cause infectious plant diseases such as fungi, bacteria, viruses, viroids, mollicutes, parasitic higher plants, and protozoa. The infectious disease means the ability of phytopathogen to transfer from the infected plant to another healthy one and causes the same disease and the same symptoms. The most phytopathogens can inhabit the internal environment of plants; however, some others can live on the plant surface such as some fungi, bacteria, and parasitic higher plants.

- ***Diseases caused by parasitic higher plants:*** Some plant diseases are developed due to growing certain plants attached on or in other plants, where they take all required nutrients without benefit sharing; these plants are called parasitic higher plants. This abnormal relationship leads to weakness of healthy or host plant. The parasitic higher plants are usually found attached with the surface of the host plant, such as dodder, mistletoe, witchweed, and broomrape.
- ***Diseases caused by nematodes:*** The nematodes are one of most common phytopathogens which have definite symptoms. These symptoms only appeared in the infected site. The nematode infections in or on plants are widely distributed especially in proper environments such as moderate temperature and high humidity.
- ***Diseases caused by fungi:*** Interestingly, there are two main types of fungi appearing on plants: pathogenic and saprophytic. The pathogenic fungi live in or on plant tissues and cause serious complications for the vital physiological functions of plants, while saprophytic ones live in or on dead tissues. Accordingly, the diagnosis of plant disease must be exactly carried out. The exact diagnosis and determination of fungi take place by microscopical examination to identify the mycelial morphological characteristics, whatever fruiting structures and spores. After complete identification for the fungus and the symptoms of plant disease, the latter should be compared with that reported in the reference. This study will exactly determine whether the fungus is a pathogen or a saprophyte. Although microscopical examination is an essential and effective method for fungal identification, it only sometimes cannot lead to exact identification due to the absence of fungal fruiting structures and spores on infected plant tissue. Therefore, an alternative method must be used, such as using selective media for isolation, identification, or promotion of sporulation. On the other

hand, some fungi need to be incubated under certain temperature, aeration, or light conditions to produce spores.

- ***Diseases caused by bacteria and mollicutes:*** The appearance of bacterial growth in or on plant tissues means that bacterial plant disease may be present, because saprophytes may be present. Therefore, accurate bacterial identification must be carried out by using microscopical examination and physiological parameter determination. The selective media are essentially used in the bacterial identification to determine the bacterial genus and species in some cases. Moreover, the confirmatory test of bacterial pathogenicity may be carried out by inoculation of single pure bacterial colony in the healthy plant, reproducing the same symptoms that appeared on the infected one. Moreover, immunodiagnostic techniques or serodiagnostic assays can be used, such as agglutination and precipitation, fluorescent antibody staining, and enzyme-linked immunosorbent assay (ELISA). There are several advantages for these techniques such as quite sensitivity, fairly specificity, rapid, easy to perform, and it is expected that standardized, reliable antisera will be available soon. Furthermore, there are recent methods used for bacterial identification, which depend on the automated analysis of bacterial fatty acid profile. The molecular biological techniques are also widely used. There are uncommon microorganisms called mollicutes. These microorganisms are very small where they must be examined by an electron microscope. Mollicutes have polymorphism and lack cell wall–like mycoplasma. These microorganisms habit the young phloem cells as a convenient host, and cause severe plant diseases such as plant stunting, yellowing or reddening of leaves, proliferation of shoots and roots, production of abnormal flowers, and eventual decline and death of the plant. Mollicutes cannot be cultured on nutrient media except for the genus *Spiroplasma*. Mollicutes can be diagnosed by several parameters, such as symptoms determination, grafting, transformation, microscopical examination, susceptibility to tetracyclines, and so on.
- ***Diseases caused by viruses and viroids:*** There are distinctive types of plant diseases caused by viruses/viroids. These diseases have definite and clear symptoms, which easily support disease diagnosis and are considered main advantage. Apart from this advantage, some recent techniques are widely used for disease diagnosis and virus identification, such as virus transmission tests to specific host plants by sap inoculation, grafting, certain insect, nematode, fungus, and mite vectors. Moreover, serodiagnostic tests are used for this purpose such as enzyme-linked immunosorbent assays, gel diffusion tests, micro-precipitin tests, and fluorescent antibody staining. The electron microscopy techniques as negative staining of virus particles in leaf

dip or purified preparations are also used, as well as immune-specific electron microscopy. On the other hand, there are more accurate techniques used for disease diagnosis and virus/viroid identification, such as electrophoretic tests and hybridization of commercially available radioactive DNA complementary to a certain virus DNA or RNA, or viroid RNA, with the DNA or RNA present in plant sap and attached to a membrane filter (immunoblot).

- ☆ ***Diseases caused by more than one pathogen:*** Sometimes, some plants are exposed to coinfection by two or more pathogens, which lead to the same or different disease symptoms. Therefore, the differentiation and identification of these pathogens are very essential to exactly determine the disease cause. The differentiation and subsequently identification are carried out by all techniques that are mentioned above.

Noninfectious Plant Diseases

Occasionally, some plant diseases have abiotic origin such as environmental factors; these diseases are called noninfectious diseases. Abiotic environmental factors have deleterious effects on plants under extreme conditions, because they can negatively effect on the vital physiological functions and may lead to death, for example, the presence of considerable amounts of toxics in the soil or in the air, deficiency of water, oxygen, or minerals, and extreme conditions for temperature, humidity, oxygen, CO, or light.

Parasitism and Pathogenicity

The term parasitism called on the state in which an organism (parasite) lives on or in another one (host) to obtain its required nutrition. Usually, the parasitism is correlated with pathogenicity, which means the ability of an organism to cause a disease. However, the parasitism in some cases leads to a benefit relationship called symbiosis, in which both plant and organism alternate the benefits, such as bacterial nodules in the roots of legume plants and the mycorrhizal infection of feeder roots of most flowering plants.

In the case of parasitism-pathogenicity relationship, the plant is diseased with the appearance of different symptoms such as increased respiration, disintegration or collapse of cells, wilting, abscission, abnormal cell division and enlargement, and degeneration of specific components such as chlorophyll. The most common plant pathogens are fungi, bacteria, mollicutes, parasitic higher plants, parasitic green algae, nematodes, protozoa, viruses, and viroids. These parasites cause serious plant diseases, because they have the ability to penetrate the plant tissues to feed and proliferate in it, and withstand the conditions in which the host lives. These pathogens are also called obligate parasites because they can only live in their living hosts. On the other hand, there are certain pathogens such as most fungi and bacteria can live on either living or dead hosts and on various nutrient media, so they are called nonobligatory parasites.

Some nonobligatory parasites can grow saprophytically on dead organic matter, and therefore called semi-biotrophs/facultative saprophytes. There is a type of a life called facultative parasitism, in which an organism grows saprophytically (necrotrophs); however, under certain conditions, they attack living plants and cause a disease; these parasites are called facultative parasites. The type or degree of parasitism does not affect the disease severity. For instance, many diseases caused by weakly parasitic pathogens are much more damaging to a plant than others caused even by obligate parasites. Lysozymes are a main mechanism of most nonobligatory parasites by which they can degrade the plant cell wall and subsequently cause invasion and infection.

Host Range of Pathogens

Phytopathogens differ among each other with respect to the plant type, the location of infection, and the age of the organ or tissue (location of infection). The specificity of plant pathogens has various degrees; some pathogens have only one target species of plant, while other pathogens can attack only one genus of plants, and eventually some others have a wide range of hosts, belonging to many families of higher plants.

As mentioned above, phytopathogens differ among each other with respect to the location of infection; some of them grow on roots, stems, leaves, fruits or vegetables, and phloem or xylem. Some phytopathogens can only infect the seedlings or the young parts of plants, while the others can only infect the mature tissues.

Development of Plant Disease

The plant disease means the occurrence of physiological disorder(s) due to biotic agents such as microbial infection and/or abiotic agents such as extreme environmental factors. In order for the plant disease to occur, an interaction must happen between two components: the plant and disease cause, which leads to physiological disorders. The disease cause is either biotic agent or abiotic agent as mentioned above. Interestingly, the biotic agents lead to infectious diseases, which develop under suitable environmental conditions. Therefore, the infectious diseases (occurred by pathogens) are not developed under extreme environmental conditions. This means it was impossible to get infectious and noninfectious plant diseases at the same time.

The abiotic agents (environmental factors) play an important and vital role in the disease development and severity or disease resistance. This matter depends mainly on different factors: the plant family, the plant age, plant genetic type, pathogen virulence race, pathogen inoculum size, and pathogen dormant state. Therefore, we can imagine the plant disease as a triangle, which is called "disease triangle." The three sides of this triangle are the plant, microorganisms, and the environmental factors. The length of each side is proportional to the sum of the characteristics of the other two sides. For example, if the plant is resistant, the host side and the amount of disease would be small or zero, whereas if the plant is susceptible, the host side would be long and the potential amount of disease could be great.

Plant Disease Cycle

The disease cycle is a series of definite events, which lead to the disease development and pathogen propagation. These events include inoculation, prepenetration, penetration, infection, colonization (invasion), and growth and reproduction of the pathogen.

Inoculation

Inoculation is the pathogen or any part of the pathogen that contacts with the plant at certain site to initiate the infection process, such as spores, sclerotia, or fragments of mycelium of fungi may be fungal inoculum. In some cases, the inoculum is represented as an intact cell as in bacteria, mollicutes, protozoa, viruses, and viroids. There are two types of inoculum: primary and secondary inoculum, which in turn cause primary and secondary infection. The primary inoculum lives dormant in the winter or summer and causes the original infections in the spring or in the autumn. The secondary inoculum is that produced from primary infections. The primary inoculum is more abundant than secondary inoculum and closer to the crop, and caused more severe diseases and the losses that result. The inoculum has two sources: inside and outside sources. The inside source in which the inoculum is produced on the plant, plant debris, or on the soil, such as fungal and bacterial inocula of perennial plants, is produced on the branches, trunks, or roots of the plants. The outside source of inoculum is in which the inoculum comes into the field with the seed, transplants, tubers, or other propagative organs or it may come from sources outside the field. In some cases, the inoculum is produced on the plant surface as in fungi, bacteria, parasitic higher plants, and nematodes, which either produce their inoculum on the surface of infected plants or their inoculum reaches the plant surface when the infected tissue breaks down. However, the inoculum may be produced within the plant as in viruses, viroids, mollicutes, fastidious bacteria, and protozoa. Interestingly, there is an expression called inoculum landing or inoculum arrival, which means incoming of the inoculum to the host plants passively by wind, water, and insects.

Prepenetration

Attachment of Pathogen to Host

Some pathogens directly penetrate the plant tissues by their vectors and then are surrounded by cytoplasm, cell membrane, or cell wall of plant cell, such as mollicutes, fastidious bacteria, protozoa, and most viruses. In other cases, the pathogen firstly makes contact with the external surface of the plant, and then penetration process takes place, such as fungi, bacteria, and parasitic higher plants. The adhesion of the pathogen with plant surface is carried out by mucilaginous substances found on the pathogen surface or at its tip. These substances are composed of mixture of water-insoluble polysaccharides, glycoproteins, lipids, and fibrillary materials, which, when moistened, become sticky and help the pathogen adhere to the plant. In some fungi as powdery mildew, adhesion is carried out by the release of cutinase enzyme from the spore, which makes the plant and spore areas of attachment more hydrophilic and cements the spore to the plant surface.

Spore Germination

Spore germination process initiates by growth stimulation, which takes place with the availability of proper environmental conditions. Once the stimulation has been received, the spore starts to utilize the stored food, such as lipids, polyoles, and carbohydrates to build germ tube as abridge with cell membrane and cell wall of the plant. When appropriate physical and chemical signals, such as surface hardness, hydrophobicity, surface topography, and plant signals, are present, germ tube extension and differentiation take place.

Appressorium Formation and Maturation

Appressorium is a specialized cell typical to many fungal plant pathogens that is used to infect the plant host. Once appressoria are formed, they adhere tightly to the leaf surface and then penetrate the plant cell wall via lysozyme secretion.

Recognition between Host and Pathogen

When a pathogen comes in contact with a host cell, the plant triggers a signal that either allows or retards the pathogen growth and development of disease. This signal is a biochemical reaction, which acts as a receptor to a pathogen contact. The pathogen propagation depends on the components of the plant cell, such as fatty acids galacturonan, phenolic compounds, strigol, amino acids, and sugars.

- **Spores and seed germination:** The availability and ability of host infection are increased by vegetative pathogen. The infection by fungal spores or parasitic higher plant seeds is carried out after germination has achieved. Fungal spores' germination is carried out by releasing either a mycelium or a germ tube that grows into the plant cell and cause infection.
- **Growth of nematodes:** The growth of nematodes starts with hatching of eggs, which essentially requires convenient environmental conditions such as temperature and moisture. After hatching of the eggs, the larvae penetrate the plant cell and grow to form the adults. After maturation, the adults of nematodes closely adhere with plant roots due to some factors, such as carbon dioxide and amino acids.

- **Penetration:** Phytopathogens penetrate plant surfaces either through natural openings such as fungi and nematodes or through wounds in cell wall such as bacteria, viruses, viroids, mollicutes, fastidious bacteria, and protozoa. Penetration and infection are not usually correlated together, because some penetrated plants are resistant to phytopathogens.
- **Infection:** The intimate contact of phytopathogen with its host is called infection process. The infection process is either successful or unsuccessful depending on the type of host, whether susceptible or resistant, respectively. Successful infection results in the appearance of symptoms, such as discoloration, necrosis, dwarfism, and so on of the host. While unsuccessful (latent) infection does not lead to any observations for the symptoms. As well known, the symptoms start to appear after the incubation period of the pathogen has been finished. The symptoms either are stable or may be changed until death.
- **Invasion:** The phytopathogens can invade the plant tissues by producing mycelia which grow between the cuticle and epidermis, such as pathogenic fungi of an apple. Nevertheless, other phytopathogens such as those causing powdery mildews produce mycelia which grow on the plant surface, and then extend to form a structure called haustoria, which in turn extend into the epidermal cells. Therefore, plant pathogenic fungi can invade their host either by intracellular mycelia, which directly grow through the cells, or by intercellular mycelia, which grow between the cells. On the other hand, plant pathogenic bacteria invade the plant tissues via intercellular way, and cause vascular wilts. Although nematodes can invade the plant tissues intercellulary or intracellularly, they usually feed on the epidermal cells by piercing. Other phytopathogens such as viruses, viroids, mollicutes, fastidious bacteria, and protozoa can intracellularly invade the plant tissues.
- **Growth and reproduction of the pathogen:** Most phytopathogens especially fungi and parasitic higher plants invade and infect plant tissues through the point of inoculation. Therefore, these pathogens can easily grow and spread within the plant tissues until a certain limit or death occurs. For example, fungi can invade and infect the plant tissue to cause vascular wilts. This invasion is carried out by releasing spores within the vessels.

Laboratory Tests of Infectious Plant Disease

Sometimes neither symptoms nor signs provide enough specific or characteristic information to decide the cause of an infectious plant disease. In such cases, it may be necessary to bring a sample back to the laboratory for further tests to isolate and identify the causal agent. This can be a time-consuming and labor-intensive process that takes specialized skills.

Incubation of Plant Material

One of the first steps when getting back to the laboratory may be to place a sample of the diseased tissue under conditions that will allow an infectious agent to grow and possibly induce sporulation. This can be accomplished by placing a leaf in a moist chamber. A moist chamber can be a sterile petri dish containing a wet filter paper in the bottom of the dish and a triangle of glass tubing on which the sample is placed so that the sample is not directly on the wet filter paper but is exposed to humid conditions. This type of moist chamber will work for small and relatively flat specimens such as leaves. Plastic bags or boxes may be necessary for larger specimens. Saprophytes that are present on the specimen can also be encouraged to grow in a moist chamber and a brief surface swab with 70 per cent isopropanol or 0.1-1 per cent sodium hypochlorite may be useful in reducing these saprophytes. Moist chambers are generally incubated at room temperature.

Isolation and Identification of Biotic Plant Disease Causal Agents

Isolation of fungi usually requires that pieces of infected plant tissue be placed on various nutrient media. The organism that grows out of this tissue is then isolated in pure culture. Bacteria are often isolated by chopping up infected tissue in a small amount of sterile water. This water : bacteria suspension is then streaked onto a bacteriological medium such as nutrient agar. Several problems can occur when trying to isolate the plant pathogenic agent. The infected plant tissue may contain one or more saprophytes which have moved into the infected tissue. These saprophytes may outgrow the plant pathogen on the nutrient medium, obstructing accurate identification of the pathogen. In some cases where a specific plant pathogen is suspected, a medium selective for the suspected pathogen may be utilized. It is also beneficial to attempt to isolate the plant pathogen from the margins of the diseased tissue where the pathogen is more numerous or more active than saprophytes that quickly colonize the recently killed tissue.

Once an organism is isolated, is that organism the true cause of the problem? Conducting Koch's postulates[1], which involves the inoculation of healthy plants, may be necessary to conclusively answer this question, especially if the organism has not been previously reported as a plant pathogen on that host. Koch's postulates are seldom conducted for routine diagnoses, but may be extremely important for new diseases and for research. Inoculation of a healthy host and obtaining the symptoms originally observed in the field may be difficult. This may be due to problems in replicating the conditions through which the host was inoculated and also in reproducing the environmental conditions present when the host became infected. It is often impossible to replicate in the laboratory the original conditions present during the disease development.

Once an infectious plant pathogen is successfully isolated, the organism must be identified. There are estimated to be some 1.6 million fungal species, most of which are not infectious pathogens. Many fungi and bacteria have never been isolated and identified. The characteristics upon which their identification is based are often complex and specialized training is necessary to be able to identify these fungi and bacteria. Diagnosticians with experience are often able to identify the most commonly isolated organisms. Identification of plant pathogenic nematodes also requires a trained individual.

Different Diagnostic Tests for Identification of Biotic Causal Agents

A major problem in identification of biotic causal agents is the inability of some infectious pathogens to grow on artificial media. Viruses, as well as some fungi (*e.g.* powdery and downy mildew causing agents) and some prokaryotes (*e.g.* phytoplasmas), require a living host in order to grow. In cases where the plant pathogen is difficult or impossible to grow on artificial media, other methods may be used for their detection, such as the use of serological tests for viruses. Viral identification is often accomplished utilizing ELISA (enzyme-linked immunosorbent assay) which is based on the binding of an antibody produced to a specific virus with the virus in the infected plant material[1]. More tests are currently being developed using the polymerase chain reaction (PCR) for detection of specific organisms. These types of reactions take specialized equipment and reagents, and the tests are not commonly done outside diagnostic and research laboratories. Other techniques used for the identification of viruses include negative staining and electron microscopy to view the viral particles in plant tissue or suspensions. PCR and ELISA tests, as well as other laboratory tests, may be used for organisms that will grow on artificial media. Additional tests may include analysis of fatty acids of organisms, carbohydrate utilization (*i.e.* BIOLOG test), and enzyme activity testing (*i.e.* pectinase, isozyme patterns).

Different Diagnostic Tests for Identification of Abiotic Causal Agents

It is extremely important to look for abiotic factors that may be important in observed symptoms. Soil and water tests may be necessary to determine pH, nutrient composition, salinity, and other factors such as pesticide residues that may induce various symptoms. It may also be important to get samples of plant tissue analyzed for nutrient content to determine if there are macro- or micronutrient deficiencies or toxicities.

Final Diagnosis of a Plant Disease

Diagnosis is a form of hypothesis testing, where the hypothesis is simply the identity of the disease, and a good diagnostician goes through multiple iterations of the scientific method (seeking evidence through testing that supports or refutes the hypothesis that s/he generates). These hypotheses are generated through observations of the plant, environment, and information from the grower. When all of the information is successfully collected, literature sources should be consulted to determine what is already known about diseases and disease-causing agents associated with the identified plant.

Information can be obtained from published resources including plant disease compendia, plant disease indexes, technical notes, commodity newsletters, online resources, and personal communication with plant disease experts. When no information is available on the specific plant, information on diseases and disease causing-agents of similar plants may be useful. There may also be rare cases where no information is available related to the disease. Then, extensive testing may be necessary to determine the identification of the plant pathogen. When this type of testing is required, it may take a long time to develop research-based control recommendations and control measures may have to be based on diseases of similar etiology. If these diseases have occurred in other areas of the world, control measures that have been previously developed in other areas may be useful.

The student should keep in mind that s/he is a detective. Causal agent identification and diagnosis of plant problems is just like a detective investigating an assault or murder case, only in this case, the victim is a plant. All clues should be investigated. Some clues may lead down blind alleys while others will lead down the correct path. It is important to note that there are exceptions that exist and these exceptions must be considered. It is the compilation of the information and clues that will ultimately lead to the most accurate diagnosis.

The figures that follow illustrate some of the common symptoms that may be produced by different types of problems. Images such as these symptoms and signs are often used in the diagnostic process. Studying these images may assist the diagnostician in narrowing down the possible diseases to consider and others which can be eliminated.

Commonly Observed Symptoms and Signs of Plant Diseases

Fungal Leaf Spots

Spots usually vary in size. Generally are round and occasionally elongate on stems. Zones of different color or texture may develop giving the spot a "bull's eye" effect. Spots are not limited by leaf veins (Figure 6.1).

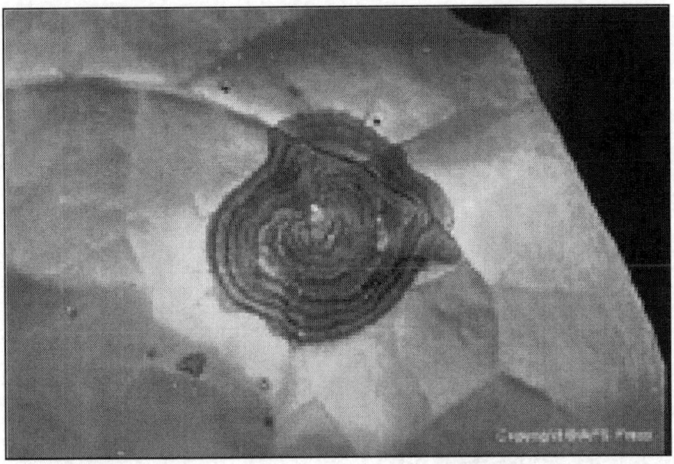

Figure 6.1: Target Spot Lesion on Tobacco Caused by *Rhizoctonia solani* (Courtesy H.D. Shew and T.A. Melton).

Bacterial Leaf Spots

Spots are often angular due to limitation by leaf veins. Color is usually uniform and no signs of plant pathogen are evident. Tissue may appear initially as being water soaked but may become papery as it dries (Figure 6.2).

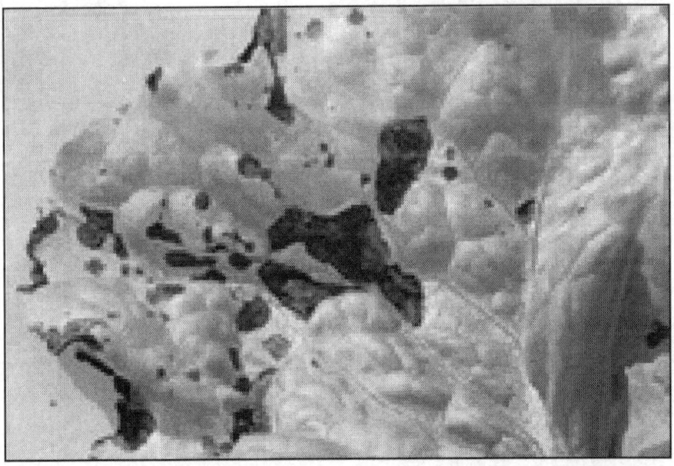

Figure 6.2: Bacterial Leaf Spot on Greenleaf Lettuce Caused by *Xanthomonas campestris* pv. *vitians* (Courtesy S.T. Koike).

Vein Banding

Vein banding occurs when there is a band of yellow tissue along the larger veins of the leaf. This symptom is observed with viral diseases and is in contrast with nutrient deficiencies which may cause a dark green band along leaf veins (Figure 6.3).

Figure 6.3: Pea Infected with Red Clover Vein Mosaic Virus Exhibiting a Vein Chlorosis and Banding (Courtesy: R.O. Hampton).

Mosaic and Ringspot

Mosaic (Figure 6.4) and ringspot (Figure 6.5) are used to describe an irregular patchwork of green and yellow areas over the surface of a leaf. In some cases leaves may also become distorted. Often these symptoms are associated with viral pathogens. There is not a sharp margin between the affected and healthy areas. Distinct margins may indicate a nutritional problem or genetic variegation.

Figure 6.4: Mosaic Symptoms Exhibited on Leaves of Squash (Used by permission of M. Riley, from files of W. Witcher).

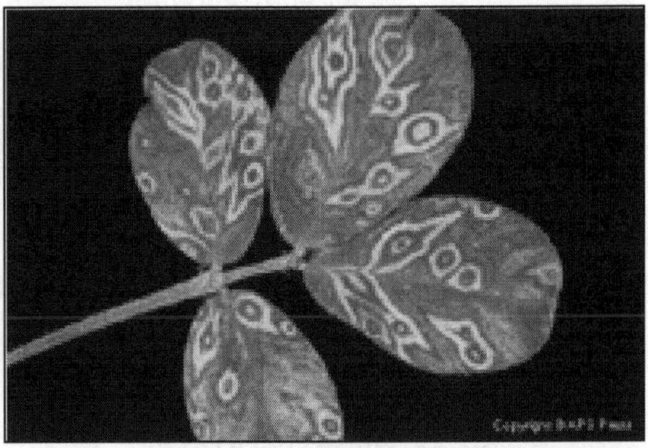

Figure 6.5: Peanut Leaf with Concentric Ring Spots Caused by *Tomato spotted wilt virus* (TSWV) (Courtesy A. Culbreath, J. Todd, and H. Pilcher).

Leaf Distortion

Leaves of the infected plant may be distorted from their normal shape and size. Leaves may be elongated, smaller size, or thickened. This type of symptom can be associated with viral, fungal or bacterial infections (Figure 6.6) as well as insect and mite infestations.

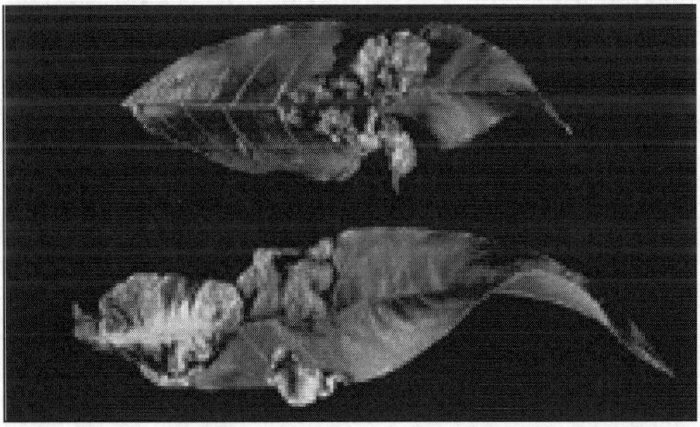

Figure 6.6: Peach Leaf Curl Caused by *Taphrina deformans* (Courtesy J. W. Pscheidt).

Powdery Mildew

It can affect leaves, stems, flowers and fruits with a white to gray surface coating of mycelia which can be rubbed off (Figure 6.7). Black specks may later develop in the mycelia. These specks are mature cleistothecia, the overwintering fungal structures which contain ascospores. Tissue may turn yellow, reddish or remain green under the mycelia and some leaf distortion may be observed especially on actively growing tissues.

Figure 6.7: Powdery Mildew on Apple Blossom Cluster Caused by *Podospaera leucotricha* (Courtesy K.D. Hickey)

Presence of Spores/Spore Structures

Several fungal diseases can be easily identified based on the presence of spores or spore structures on the leaf surface. Some examples of this are rusts which are often recognized by the rusty brown to black spores (Figure 6.8) and smuts which are identified by the black spores which often replace the seed structure (Figure 6.9).

Figure 6.8: Stem Rust on Barley. Caused by *Puccinia graminis* (Courtesy B. Steffenson).

Figure 6.9: Loose Smut (on left) and False Loose Smut (on right) on Barley, Caused by *Ustilago nuda* (left) and *U. nigra* (right) (Courtesy P. Thomas).

Needle Drop in Conifers

Conifers normally retain their needles for several years but these needles will eventually be lost. This drop is gradual and production of new needles obscures the loss of older needles. Unfavorable growing conditions, such as drought, may cause an acceleration of needle drop. If the drop occurs in only older needles especially during unfavorable growing conditions, there is no need for concern. If new needles are lost then other factors may be involved such as needle cast fungus (Figure 9.10), nutrient deficiencies, or toxic chemicals.

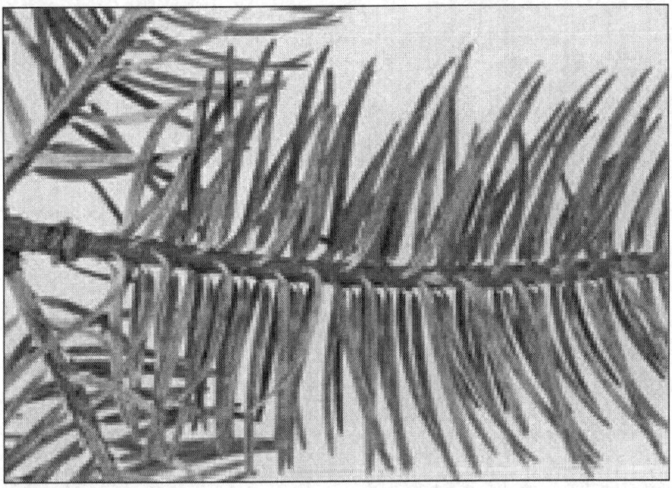

Figure 9.10: Rhabdocline Needle Cast on Douglas Fir (Courtesy E. Hansen).

Chemical Spray or Air Pollutant Injury

Spots associated with injury are relatively uniform in color and the interface between the affected and healthy area is usually sharp. Distribution on plant may be associated with where spray or pollutant comes in contact with the plant (Figure 9.11).

Leaf or Needle Tip Death

Death at the tips of needles and tips and margins of leaves often indicate unfavorable climatic conditions, toxic chemicals or root malfunction due to poor cultural practices. Air pollutants, soil chemicals, and excess fertilizer can cause tip burn. Drought and freezing may have a similar effect. All needles of a specific growth period are usually affected. Needles infected by foliar fungal diseases are generally more scattered and rarely are all needles of a particular growth period killed (Figure 9.12). Needles diseased by infectious agents are generally affected over varying lengths and are often straw yellow or light tan. Fungal fruiting bodies may also be observed.

Figure 9.11: Injury of Beans Caused by Drift of the Herbicide, Paraquat (Courtesy H. F. Schwartz).

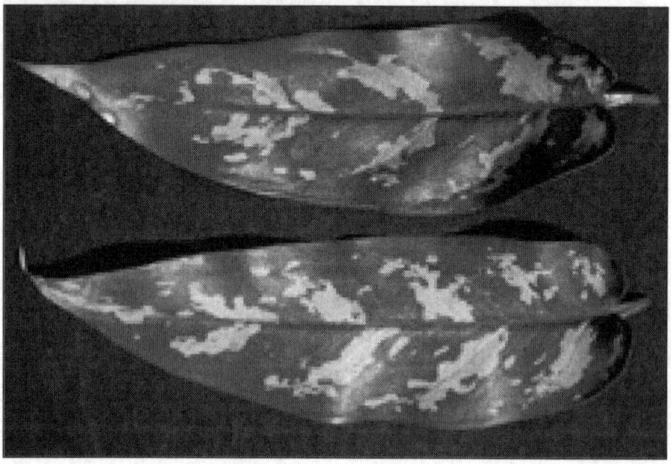

Figure 9.12: Chlorotic and Necrotic Lesions on Leaf Tips Due to Fluoride Toxicity (*Aglaonema commutatum* 'Maria') (Courtesy J. M. F. Yuen).

Soil or Air Chemical Injury

Chemicals which are absorbed from the soil by roots or absorbed from air through leaves may exhibit a burning or scorching of leaf margins (Figure 9.13). If severe, islands of tissue between the veins may also be killed. Dead tissue may drop out of the leaf leaving a ragged appearance. Other chemicals may cause a distortion of leaf shape and size.

Figure 9.13: Alfalfa Leaves Exhibiting Scorching of the Leaf Edges due to Sulfur Dioxide Injury (Courtesy K. T. Leath).

Cankers

Cankers are localized necrotic lesions which are often sunken in appearance (Figure 9.14). Cankers can result from mechanical injury (*e.g.* trees which have been damaged by collisions with cars or lawnmowers), and various fungi or bacteria. In the spring, ooze may be observed on the surface of bacterial cankers and fruiting bodies may be observed on the surface of fungal cankers.

Fruit Decays and Rots

Various fungi and bacteria can cause rots of fruit. These are often distinguished by the color, lack of firmness of tissue, and signs of spores or fruiting bodies (Figure 9.15).

Figure 9.14: Canker on Apple Caused by *Nectria galligena* (Courtesy A. L. Jones).

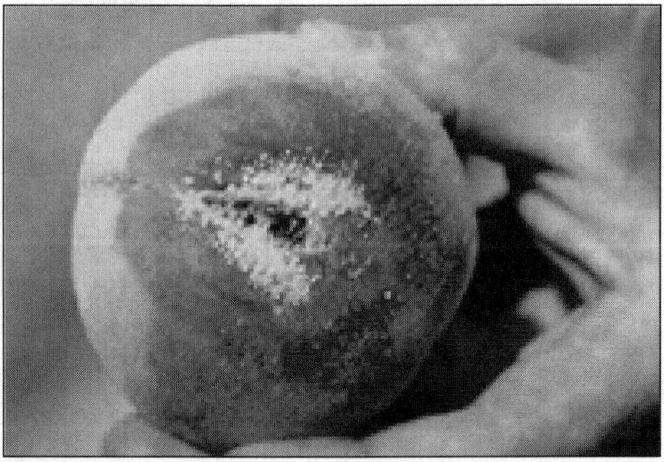

Figure 9.15: Brown Rot of Peach Caused by *Monilinia fructicola* (Courtesy J. M. Ogawa).

Fruit Discoloration

Discoloration of fruit is often associated with viral infections (Figure 9.16). This discoloration may be similar to mosaic and ringspot symptoms observed on leaves (Figure 9.17).

Figure 9.16: Mosaic Symptoms on Yellow Crookneck Squash. Normal appearance is completely yellow fruit. (Used by permission of M. Williamson)

Figure 9.17: Ringspot Symptoms on Peach Fruit due to *Plum pox virus* (Courtesy A. N. Adams).

Wilts

Wilts are characterized by a general loss of turgidity of leaves or possibly entire plants due to the loss of water (Figure 9.18). The loss is most often caused by a blocking of the water flow through the xylem. This blockage can be caused by the presence of various bacteria (*Erwinia, Ralstonia*) and fungi

(*Fusarium, Verticillium*) in the xylem. Wilts may also be observed when there is a destruction of the root system due to nematodes or root-rotting fungi (*Armillaria, Phytophthora, Pythium*) or an acute water shortage in the soil.

Figure 9.18: Verticillium Wilt of Cucumber Caused by *Verticillium dahliae* Courtesy W. D. Gubler).

Shoot Dieback or Blights

Sudden dieback of a shoot usually indicates climatic or chemical injury rather than parasitic problems. If the line between affected and healthy bark is sharp, a soil chemical should be suspected. If dieback is somewhat more gradual and there is a cracking or splitting of the bark and wood, cold injury should be suspected along with bacterial blights caused by *Pseudomonas* or *Erwinia*. A

Figure 9.19: Bacterial Ooze from Cut Tomato Stem Infected with *Ralstonia solanacearum* (Used by permission of M. Williamson).

bacterial streaming test with phase and compound microscope and isolation may be required to determine if the blights/dieback is due to a bacterial agent. Gradual decline of shoots and retention of dead leaves is more indicative of a parasitic disease. The margin between the affected and healthy tissue is often irregular and sunken. There may also be small pin-like projections or bumps over the surface of dead bark. These bumps are spore-producing structures of the fungal causal agent.

Figure 9.20: Shoot Growth of Apple with Fire Blight, *Erwinia amylovora*, Showing Typical Shepherd's-Crook at the Tip of the Shoot (Courtesy A. L. Jones).

Dying Branches of Trees and Shrubs

If scattered branches in a tree or shrub start to decline and eventually die, a canker disease or a shoot blight should be suspected (Figure 9.21). If branches die suddenly and, especially if affected branches are concentrated on one side of tree, weather conditions should be suspected (wind, snow, *etc.*) or animal or mechanical damage at the trunk base; however, this is not always the case. If symptoms develop over time on one side of a tree or plant then damage of the roots associated with one side may be suspected such as root rots due to *Phytophthora* spp.

Death of Tree and Shrub Top

If all or a major portion of a tree or shrub dies over a period of time, the diagnostician should suspect a problem with the roots (Figure 9.22). Examples are diseases caused by Armillaria root rot and Verticillium wilt. The decline may be gradual and may eventually affect the entire tree, but in some cases the death may occur on one side of the plant initially. If the decline is sudden, a toxic chemical in soil or weather extremes such as freezing or drought should be suspected.

Figure 9.21: Branch Dieback of Monterey Pine in California Caused by Pitch Canker (Courtesy L. D. Dwinell).

Figure 9.22: Crown Symptoms Caused by Laminated Root Rot of Douglas-fir (Courtesy W. G. Thies).

Overall Stunting or Decline

These symptoms can be caused by several very different factors. Systemic viral infections can result in stunting or decline, but such viral infections are often accompanied by other aboveground symptoms such as shortened internodes. In many cases, overall stunting of a plant may be due to problems associated with the root system (Figure 9.23). The roots should be examined for rotting and possible mycelial growth, reduction in roots especially feeder

Figure 9.23: Healthy 'Mandalay' Chrysanthemum (left) and Plant Infected with *Fusarium oxysporum* f. sp. *chrysanthemi* (right) which Exhibits Stunting without other Observable Symptoms (Courtesy Penn State Univ.).

Figure 9.24: Crown Gall on Euonymous Caused by *Agrobacterium tumefaciens* (Courtesy Robert L. Forster).

roots, and the presence of galls. Root galls can result from fungal and fungallike agents (*Plasmodiophora brassicae)*, nematodes (*Meloidogyne* spp. - root-knot), and bacteria (*Agrobacterium* sp.). Abiotic factors such as nutritional deficiencies, soil compaction and herbicide residues can also result in overall stunting or decline.

Damping-off

This term describes the rapid death and collapse of young seedlings. Often the seedlings will appear to be almost broken at the soil line It may be observed in flats of plants begun in greenhouses and can result from infection of the seedlings by the fungal organisms *Fusarium, Phytophthora, Pythium, Rhizoctonia,* or *Thielaviopsis* (Figure 9.25).

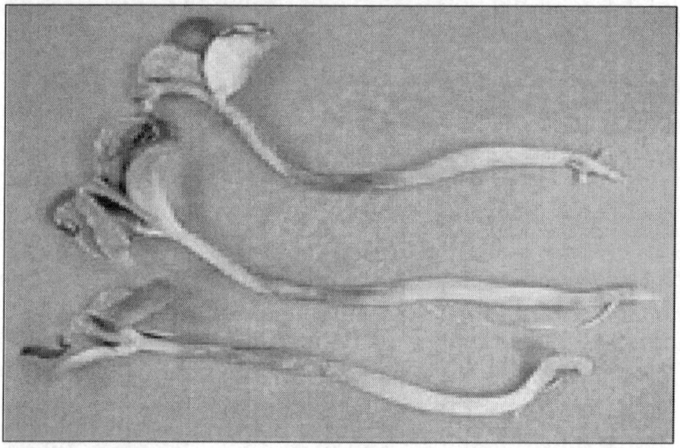

Figure 9.25: Soybean Seedlings Exhibiting Symptoms of Pythium Damping-off. Note thinning and browning of stem growth near the soil line (Courtesy X. M. Yang).

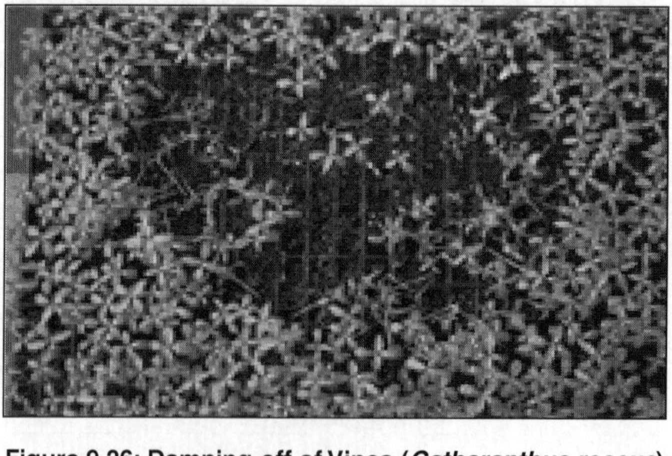

Figure 9.26: Damping-off of Vinca (*Catharanthus roseus*) due to *Rhizoctonia solani* (Courtesy R. L. Wick).

Chapter 7

Phytopathogenic Bacteria and Mollicutes

Introduction

Phytopathogenic bacteria are a group of microorganisms that cause diseases in plants. These bacteria have evolved various mechanisms to invade, colonize, and damage plant tissues, leading to a wide range of plant diseases. The study of phytopathogenic bacteria is essential in agriculture and plant pathology to understand disease development, devise control strategies, and ensure crop health.

Phytopathogenic bacteria affect all kinds of plants and cause several kinds of diseases (see 'Symptoms of bacterial infection on plants'). The range of plant species susceptible to bacterial infection includes the majority of agriculturally important crops.

The Examples of Phytopathogenic Bacteria

Most plant pathogenic bacteria belong to the genera: *Erwinia, Pectobacterium, Pantoea, Agrobacterium, Pseudomonas, Ralstonia, Burkholderia, Acidovorax, Xanthomonas, Clavibacter, Streptomyces, Xylella, Spiroplasma,* and *Phytoplasma.*

General Characteristics of Phytopathogenic Bacteria

- ✰ Bacteria are small: 0.6-3.5 microns in length, 0.5-1.0 microns in diameter
- ✰ Most plant pathogenic bacteria are facultative saprophytes and can be grown artificially on nutrient media; however, fastidious vascular bacteria are difficult to grow in culture and some of them have yet to be grown in culture.

- ☆ Bacteria may be rod shaped, spherical, spiral, or filamentous (threadlike).
- ☆ Mostly flagellate hence motile. Some bacteria can move through liquid media by means of flagella, whereas others have no flagella and cannot move themselves.
- ☆ Some can transform themselves into spores, and the filamentous bacteria Streptomyces can produce spores, called conidia, at the end of the filament. Other bacteria, however, do not produce any spores.
- ☆ The vegetative stages of most types of bacteria reproduce by simple fission.
- ☆ Bacteria multiply with astonishing rapidity they can produce tremendous numbers of cells in a short period of time. 8. Carbohydrate decomposition is mostly aerobic or oxidative (Except, Erwinia,which is facultative anaerobe)
- ☆ Mostly gram-ve, rarely gram+ve (gram+ve genera: Streptomyces, Corynebacterium, Clavibacter, Curtobacterium)
- ☆ PPB are passive invaders,*i.e.* enters plants through wounds or natural openings
- ☆ Survive in/on debris and spread by means of water,rain,insects and agricultural implements.
- ☆ Susceptible to phages,none of them cause human or human diseases.
- ☆ Cellwall is rigid
- ☆ Slow growth compared to other saprophytic bacteria.
- ☆ Incubation period: 36-48hrs 25°C

Other distinguished features;

- ☆ Their genetic material (DNA) is not enclosed within a membrane.
- ☆ They lack other membrane-bounded organelles Fig: Detailed structural composition of bacterial cell
- ☆ Their DNA is not associated with histone proteins (special chromosomal proteins found in eukaryotes).
- ☆ Their cell walls most of the time contains the complex polysaccharide peptidoglycan.
- ☆ They usually divide by binary fission. During this process, the DNA is copied and the cell splits into two cells.

Bacteria Cell

A bacteria cell is protected by cell envelop which made-up of capsule,a cellwall and plasma membrane. In some cells slime cover instead of capsule,outer plasma membrane between capsule and cellwall A variety of structures are found in prokaryotic cells. The cells are always bounded by a chemically complex cell wall. In between the cell wall and the periplasmic space lies the plasma membrane. Internal membrane bound organelles are absent in prokaryotes and hence the internal morphologically is very simple. The genetic material is localized in a discrete region called nucleoid and not separated from the surrounding cytoplasm by membrane. Ribososmes and inclusions are found scattered in the cytoplasmic matrix. Bacteria use flagella for locomotion, and in addition, many cells are surrounded by a capsule or slime layer external to the cell wall.

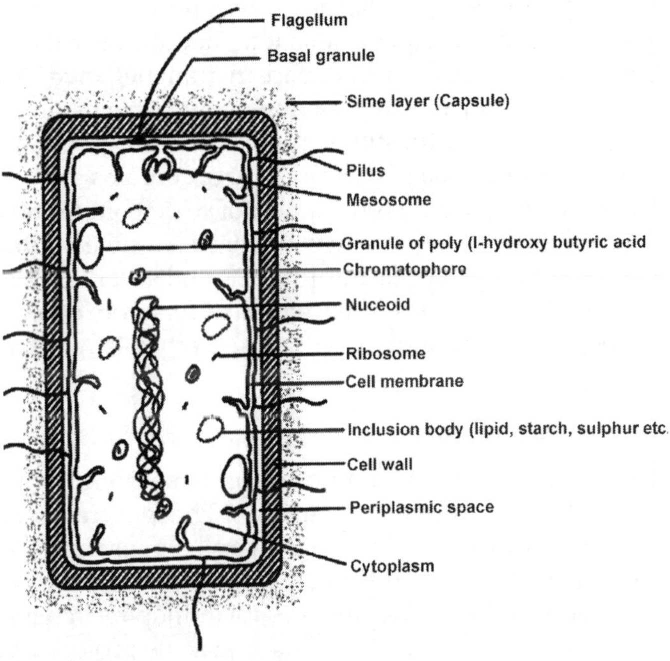

Figure 7.1: Detailed Structural Composition of Bacterial Cell.

The Bacterial Cell Wall

It's a complex, semi-rigid structure that surrounds the underlying plasma membrane and protects the cell and the internal parts from adverse surrounding environment. Except for Mycoplasmas and some Achaea most bacteria have strong walls that give them shape and protect them from osmotic lysis; wall shape and strength is primarily due to peptidoglycan. The cell walls of pathogenic bacteria have components that contribute to their pathogenicity and protect a cell from toxic substances and are the site of action of several antibiotics. Major functions – prevent cells from rupturing, maintains shape and serves as a point of anchorage for flagella. Christian Gram developed the Gram stain in 1884, and it became evident that bacteria could be divided into two major groups based on their response to staining. Gram positive bacteria stained purple and the gram negative ones colored pink or red by the technique. The gram positive cell wall consists of a single thick layer of 20 to 80 nm of peptidoglycan or murein layer lying outside the cell membrane. In gram negative cells, the peptidoglycan layer is thin of about 2 to 7 nm surrounded by a 7 to 8 nm thick outer membrane. The gram positive cells are stronger than gram negative ones because of the peptidoglycan layer.

A space is seen between the plasma membrane and the outer membrane in electron micrographs of gram negative bacteria and a similar one but smaller gap may be observed between the plasma membrane and wall in gram positive bacteria. This space is called periplasmic space. The substance that occupies the space is the periplasm. The periplasmic space in gram negative bacteria ranges from 1nm to as great as 71 nm. It contains many proteins and participates in nutrient acquisition, hydrolytic enzymes attacking nucleic acids and phosphorylated molecules. The periplasmic space contains enzymes involved in peptidoglycan synthesis and modification of toxic compounds that could harm the cell. Gram positive bacteria may not have a visible periplasmic space but appear to have as many periplasmic proteins and several enzymes that are secreted are often called exozymes. The Archaea differs from other prokaryotes. The walls lack peptidoglycan and are composed of proteins, glycoprotein or polysaccharides.

Structure of Peptidoglycan

The cell wall of bacteria is composed of a macromolecular network of polymer called peptidoglycan (murein). The polymer contains two sugar derivatives, Monosaccharides- Nacetylglucosamine (NAG) and N-acetylmuramicacid (NAM). In addition, several different amino acids, three of which – D-Glutamic acid, D-alanine, and mesodiaminopimelic acid – not found in proteins are present. The pretense of D-amino acids protects against attack by most peptidases. The backbone of this polymer is composed of alternating N-acetylglucosamine and N-acetylmuramic acid residues. A peptide chain of four alternating D- and L- amino acids is connected to the carboxyl group of N-acetylmuramic acid. Many bacteria substitute another diaminoacid, usually L-Lysine, in the third position for meso-diaminopimelic acid. Chains of linked peptidoglycan subunits are joined by cross-links between the peptides. Often the carboxyl group of the terminal D-alanine is connected directly to the amino group of diaminopimelic acid, but a peptide interbridge may be used instead. Most gram- negative cell wall peptidoglycan lacks the peptide interbridge.

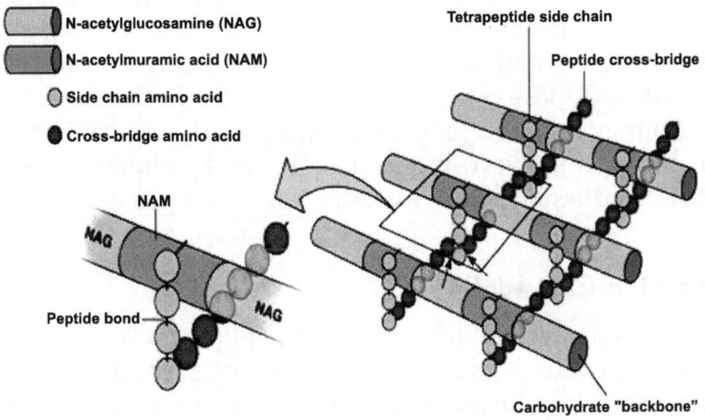

Figure 7.2: Peptidoglycan Structure in Gram-Positive Cell.

Gram Positive Cell Walls

The cell wall consists of many layers of peptidoglycan, forming a thick, rigid structure, which contains a peptide interbridge. Gram positive cell walls also contain teichoic acids, polymers of glycerol and ribitol joined by phosphate groups. To the glycerol and ribotol groups, the amino acids such as D-alanine or sugars like glucose are attached (Figure 7.3). There are two classes to teichoic acids; lipoteichoic acid (if they are attached to the lipid of the plasma membrane) and wall teichoic acid (which extend to the surface of the peptidoglycan and are negatively charged). Teichoic acids are not found in gram-negative bacteria.

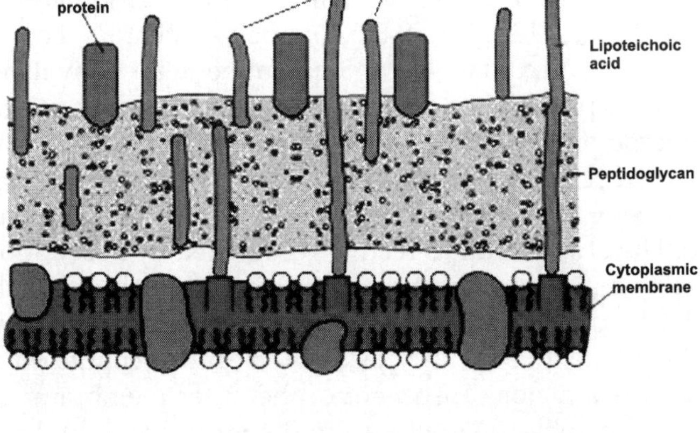

Figure 7.3: Cell Wall of Gram Positive Bacteria.

Some bacteria like Staphylococci and most other gram-positive bacteria have a layer of proteins on the surface of their cell wall peptidoglycan. Some are noncovalently attached by binding to the peptidoglycan, teichoic acids and other receptors and these proteins are involved in the interactions of the cell with its environment. Ex. S-layer proteins. Other surface proteins are covalently attached to the peptidoglycan and in gram positive pathogens these have roles such as aiding in adhesion to host tissues, preventing opsonization and blocking phagocytosis.

Gram Negative Cell Walls

The cell wall contains only a thin layer of peptidoglycan. In *E. coli,* it is about 2- 3 nmthick and contains only one or two layers of sheets of peptidoglycan. The peptidoglycan is bounded to lipoproteins in the outer membrane and is embedded in a soft material, the periplasmic gel. Gram negative cell walls do not contain teichoic acids. They are more susceptible to mechanical breakage because of small amount of peptidoglycan.

Outer membrane - found primarily in gram-negative cell consists of lipoproteins,v liposaccharides and phospholipids and lies outside the thin peptidoglycan layer. The most abundant membrane protein is Braun's lipoprotein, a small lipoprotein covalently joined to the underlying peptidoglycan and embedded in the outer membrane by its hydrophobic end. The outer membrane e and plasma membrane appear to be in direct contact at many locations in the cell wall. The adhesion sites may be regions of direct contact or possibly true membrane fusions. It is proposed that substances can move into the cell through these adhesions sites rather than travelling through the periplasm.

The most unusual constituents of the outer membrane are its lipopolysaccharides. (LPSs). These contain both lipid and carbohydrate and consist of three parts: 1. Lipid A, 2. the core polysaccharide, and 3. the O side chain. The lipid A region contains two glucosamine sugar derivatives, each with three fatty acids and phosphate or pyrophosphate attached. This region is buried in the outer membrane and the remaining portion projects from the surface. The second region or the core polysaccharide is joined to lipid A and the third region or the O side chain which is a polysaccharide chain extends outward from the core. The O side chain is readily recognized by host antibodies, gram negative bacteria may thwart host defenses by readily changing the nature of their O side chains to avoid detection. The core polysaccharide contains charged sugars and phosphate, LPS contributes to the negative charge on the bacterial surface. Lipid A is a major constituent of the outer membrane, and the LPS helps stabilize the membrane structure. Lipid A is often toxic and as a result LPS can act as an endotoxin and cause some symptoms that arise in gram-negative bacterial infections. The function of the outer membrane is to serve as a protective barrier. It prevents or slows the entry of bile salts, antibiotics, and other toxic substances that might kill or injure the bacterium. Provides a barrier to certain antibiotics (like penicillin), lysozyme detergents, heavy metals, bile salts and digestive. Porins are proteins that permit small molecules to pass through the outer membrane (about 800 MW). Specific channel proteins – allow specific substances like vitamin B12, iron, nucleotides and maltose.

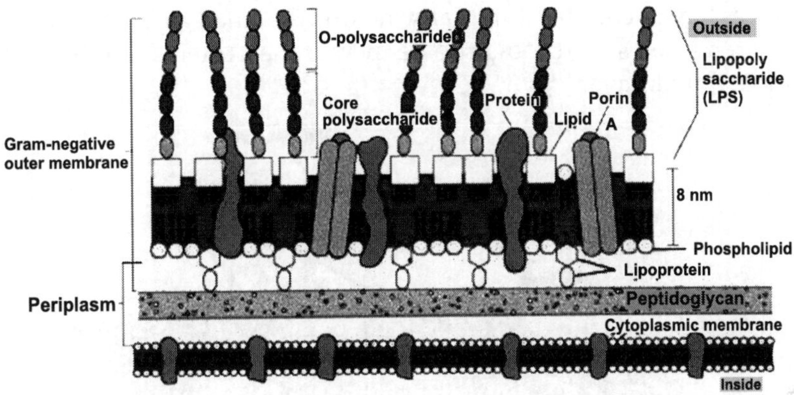

Figure 7.4: Cell Wall of Gram Negative Bacteria.

Atypical Cell Walls

Mycoplasma – is a bacterial genus that naturally lacks cell walls. Their plasma membranes have lipids called sterols, which protect them from osmotic lysis.

Archaeabacteria – have pseudomurein (N-acetylalosaminuronic acid) but no peptidoglycan.

L forms - are mutant bacteria with defective cell walls.

Structures Internal to the Cell Wall

In this lecture we shall be dealing with the structures internal to the cell wall of a bacterial cell. They include plasma membrane, organelles in the cytoplasm like nuclear area, ribosomes, inclusion bodies and endospores.

Plasma (Cytoplasmic) Membrane

Membranes are absolute requirement of all living organisms. It is the chief point of contact with the cell's environment and thus is responsible for much of its relationship with the outside world. Plasma membrane – encloses the cytoplasm and consists of phospholipids and proteins (fluid mosaic model). Most membrane-associated lipids are structurally asymmetric with polar and nonpolar ends. The polar ends interact with water and are hydrophilic and the nonpolar hydrophobic ends are insoluble in water. The lipid composition of bacterial membranes varies with environmental temperature in such a way that the membrane remains fluid during growth. Bacterial membranes usually differ from eukaryotic membranes in lacking sterols such as cholesterol and they contain pentacyclic sterol-like molecules called hopanoids and these are said to stabilize the bacterial membranes. Cell membranes are very thin structures about 5 to 10 nm thick and can be seen only with electron microscope. Plasma membranes have a complex internal structure; the small globular particles seen in these membranes are thought to be membrane proteins that lie within the membrane lipid bilayer.

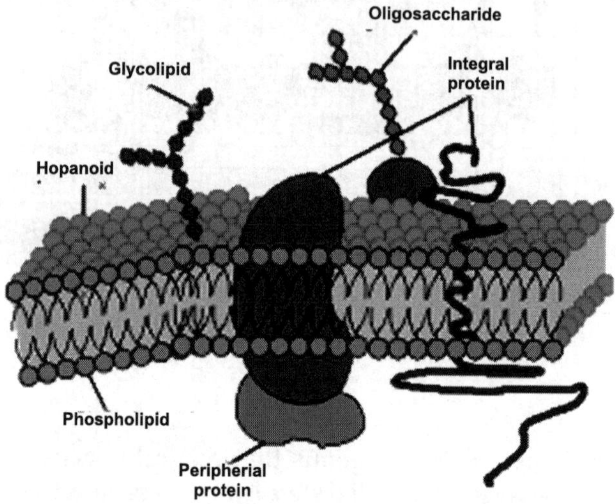

Figure 7.5: Plasma Membrane Structure.

The most widely accepted current model for membrane structure is the fluid mosaic model of S. Jonathan Singer and Garth Nicholson. Two types of membrane proteins are seen, Peripheral proteins - which are loosely connected to the membrane and can be easily removed and are soluble in aqueous solutions and make up about 20 to 30 per cent of total membrane protein. About 70 to 80 per cent of membrane proteins are integral proteins. These cannot be easily extracted from membranes and are insoluble in aqueous solutions when freed of lipids. Integral proteins, like membrane lipids are amphipathic; their hydrophobic regions are buried in the lipid while the hydrophilic portions project from the membrane surface. The plasma membrane retains the cytoplasm, particularly in cells without cell walls, and separates it from the surroundings. Plasma membranes serve as a selectively permeable barrier; it allows particular ions and molecules to pass, either into or out of the cell, while preventing the movement of others. Transport systems can be used for such tasks as nutrient uptake, waste excretion, and protein secretion. The plasma membrane also is the location of a variety of crucial metabolic processes; respiration, photosynthesis, the synthesis of lipids and cell wall constituents, and probably chromosome segregation.

The bacterial plasma membrane can be destroyed by alcohols and polymixins which cause leakage of intracellular contents and subsequent cell death of the organism.

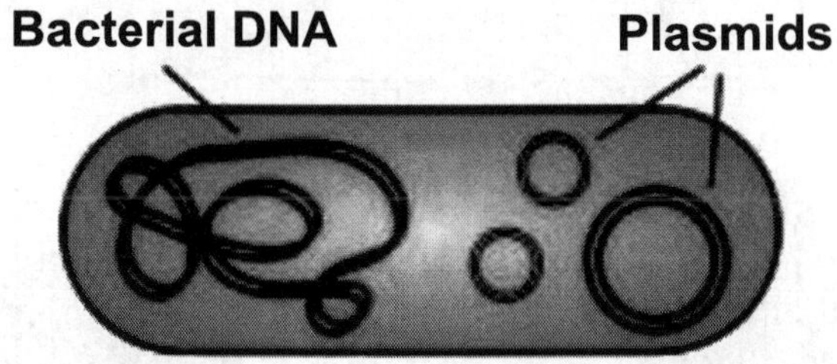

Figure 7.6: Bacterial DNA and Plasmid.

Ribosomes

The cytoplasmic matrix is also packaged with ribosomes, they also may be loosely attached to the plasma membrane. They look like small, featureless particles at low magnification electron microscope. They are made up of both protein and ribonucleic acid (RNA). Ribosome's function as sites of protein synthesis; matrix ribosomes synthesize proteins destined to remain within the cell, whereas the plasma membrane ribosomes make proteins for transport to the outside. The shape of each protein is determined by its amino acids sequence and the special proteins called molecular chaperones or chaperones aid the polypeptide in folding to its proper shape. Prokaryotic ribosomes are smaller than eukaryotic ribosomes. Ribosomes are composed of two subunits, each subunit being composed of protein and a type of RNA called ribosomal RNA (rRNA).They are comm. Only 70S: 30S subunit (1 molecule of rRNA) and 50S subunit (2 molecules of rRNA) and have dimensions of about 14 to 15 nm, a molecular weight of approximately 2.7 million. The S in 70S stands for Svedberg value or sedimentation coefficient. It is the sedimentation velocity in a centrifuge; the faster a particle travels when centrifuged, the greater is its Svedberg value. The Sedimentation coefficient is a function of a particle's molecular weight, volume and shape. Several antibiotics, such as streptomycin, neomycin and tetracyclines, exert their antimicrobial effects by inhibiting protein synthesis on ribosomes.

Inclusion bodies: Inclusion bodies can be divided into two types:

1. Inclusion bodies not bounded by a membrane and lie free in the cytoplasm. Ex. Polyphosphate granules, cyanophycin granules and some glycogen granules.
2. Inclusion bodies enclosed by a membrane about 2-4nm thick. *e.g.* Poly Bhydroxybutyrate granules, some glycogen and sulfur granules, carboxysomes andØ gas vacuoles.

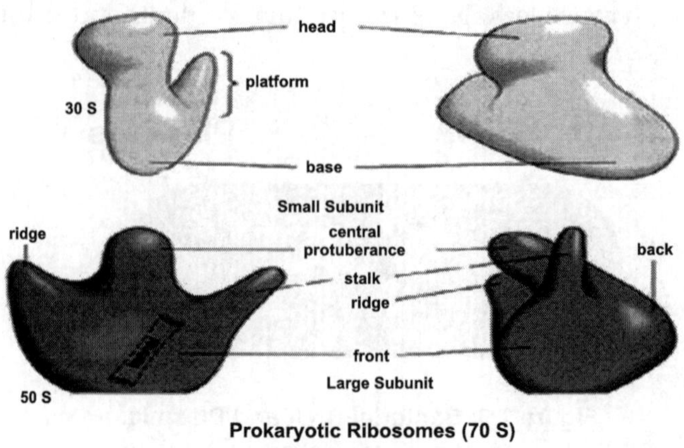

Figure 7.7: Ribosomes in Bacteria.

Organic Inclusion Bodies: Glycogen

Polymer of glucose units composed of long chains formed by alpha (1-4) glycosidic bonds and branching chains connected to them by alpha (1-6) glycosidic bonds. *e.g.* glycogen and starch, and their presence can be demonstrated when iodine is applied to the cells (glycogen granules appear reddish brown and starch granules appear blue).

Poly B- hydroxybutyrate: Contains beta-hydroxybutyrate molecules joined by ester bonds between the carboxyl and hydroxyl groups of adjacent molecules. Appear in various species of Mycobacterium, Bacillus, Azotobacter, Spirillum and other genera. Lipid inclusions are revealed by use of fat-soluble dyes, such as Sudan dyes.

- ☆ **Carboxysomes:** These are polyhedral and hexagonal inclusions that contain the enzyme ribulose 1,5- diphosphate carboxylase. ·Bacteria that use carbon dioxide as their sole source of carbon require this enzyme for carbon dioxide fixation during photosynthesis (Ex. nitrifying bacteria, cyanobacteria, and Thiobacilli).
- ☆ **Gas vacuoles:** These are hollow cavities found in many aquatic prokaryotes, including cyanobacteria, anoxygenic photosynthetic bacteria and halobacteria. Each vacuole consists of rows of several individual gas vesicles, which are hollow cylinders covered by protein. Their function is to maintain buoyancy so that the cells can remain at the depth in the water appropriate for them to receive sufficient amounts of oxygen, light and nutrients. They are impermeable to water and permeable to atmospheric gases.

Inorganic Inclusion Bodies

Polyphosphate granules or Metachromatic granules: Linear polymer of organo phosphates joined by ester bonds. Reservoirs for phosphate, an important component of cell nucleic acids and also energy reserves. Represents a reserve of inorganic phosphate (polyphosphate) that can be used in the synthesis of ATP. Stain red with certain blue dyes, such as methylene blue, and are collectively known as volutin. Found in algae, fungi and protozoans, as well as bacteria.

These granules are quite large and are characteristic of *Corynebacterium* diphtheriae, the causative agent of diphtheria, thus they have diagnostic significance.

Sulphur granules: Sulphur bacteria, which belong to the genus *Thiobacillus,* derive energy by oxidizing sulfur and sulfur containing compounds. These bacteria may deposit sulfur granules in the cell, where they serve as an energy reserve. Purple photosynthetic bacteria use H_2S as electron donor and accumulate resulting sulfurin either the periplasmic space or in special cytoplasmic globules.

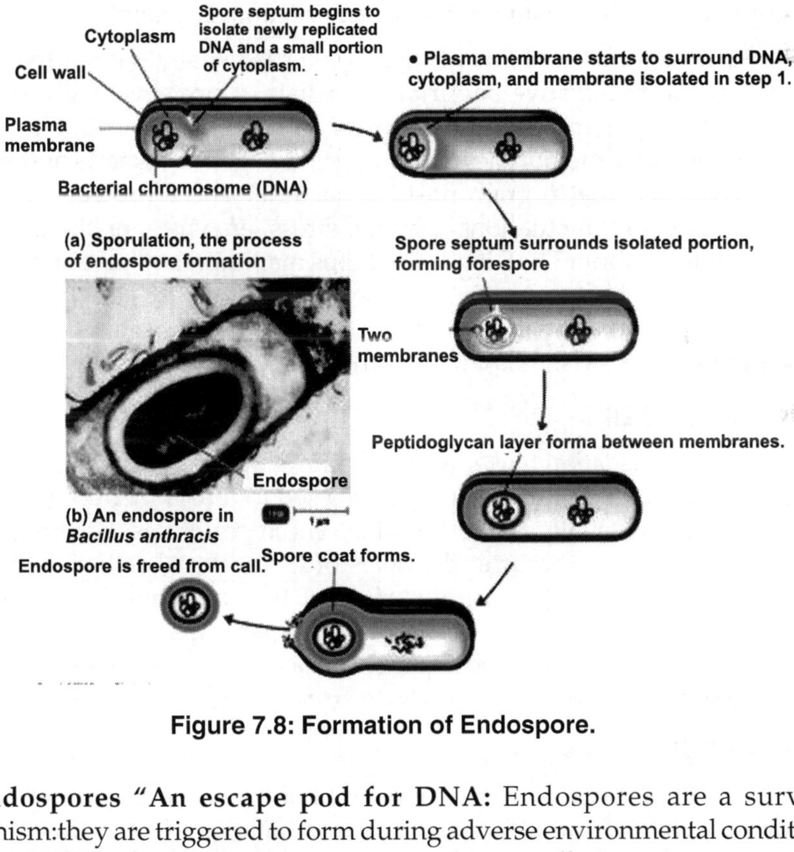

Figure 7.8: Formation of Endospore.

Endospores "An escape pod for DNA: Endospores are a survival mechanism:they are triggered to form during adverse environmental conditions. They are NOT reproductive structures as only one cell gives rise to one spore and endospores can be identified with special stains and differentiated from the vegetative cell (Figure 7.8). Endospores are resistant to: heat: withstand boiling for over one hour; desiccation, UV radiation and chemical disinfectants. The resistance of these spores has serious consequence and some very pathogenic bacteria have the ability of produce such spores.

Structures External of the Cell Wall

The structures external to the cell wall of bacterial cells includes glycocalyx, fimbriae, pili, flagella, axial filaments.

Glycocalyx (Capsules, Slime layers and S-layers)

It is a viscous (sticky), gelatinous polymer composed of polysaccharide, polypeptide or both. If the substance is organized and is firmly attached to the cell wall, the glycocalyx is described as a capsule (negative staining). If the substance is unorganized and only loosely attached to the cell wall, the glycocalyx is described as a slime layer. Capsules protect pathogenic bacteria from phagocytosis (process by which certain white blood cells engulf and destroy microbes) and contribute to virulence. Capsules are clearly visible in the light microscope by using stains or special capsule stains.

A regularly structured layer called S-layer is usually seen in many gram positive and gram negative bacteria. It consists of proteins or glycoproteins and resembles a pattern something similar to floor tiles. The S-layer adheres directly to the outer membrane in case of gram negative bacteria and with the peptidoglycan surface in gram positive bacteria. These protect the bacteria against ion and pH fluctuations, osmotic stress, enzymes, or the predacious bacterium Bdellovibrio. The S layer also helps maintain the shape and envelope rigidity of at least bacterial cells and also promotes cell adhesion to surfaces. Sometimes, the layer also seems to protect some pathogens against complement attack and phagocytosis, thus contributing to their virulence.

Fimbriae and Pili

Many gram negative bacteria have hairlike appendages that are shorter, straighter and thinner than flagella and are used for attachment rather than for motility. They are usually called fimbriae. These structures contain a protein called pilin. Fimbriae - occur at the poles of the bacterial cell, or they can be evenly distributed over the entire surface of the cell. Fimbriae of Neisseria gonorrhoeae the causative agent of gonorrhea help the microbe to colonize mucous membranes to cause the disease. At least some types of fimbriae attach bacteria to solid surfaces such as rocks in streams and host tissues.

Pilior Sex Pili or Pilus

Usually longer than fimbriae and number only one to ten per cell. Pili function to join bacterial cells prior to the transfer to DNA from one cell to another (sometimes called sex pili). They are genetically determined by sex factors or conjugative plasmids and are required for bacterial mating. Some bacterial viruses attach specifically to receptors on sex pili at the start of their reproductive cycle.

Flagella

Motile bacteria move by use of flagella, threadlike locomotor appendages extending outward from the plasma membrane and cell wall. They are slender, rigid structures, about 20 nm across and up to 15 or 20 µm long. Bacterial species often differ distinctively in their patterns of flagella distribution.

Structure

Transmission electron microscopic studies have shown that the bacterial flagellum is composed of three parts.

1. Filament– outermost region and contain the globular protein flagellin
2. Hook– the filament is attached to hook, which consists of a different protein
3. Basal body- which anchors the flagellum to the cell wall and plasma membrane. It consists of a small central rod inserted into it are a series of rings. The filament is a hollow, rigid cylinder constructed of a single protein called flagellin (MW from 30,000 to 60,000). Some bacteria have sheaths surrounding their flagella.

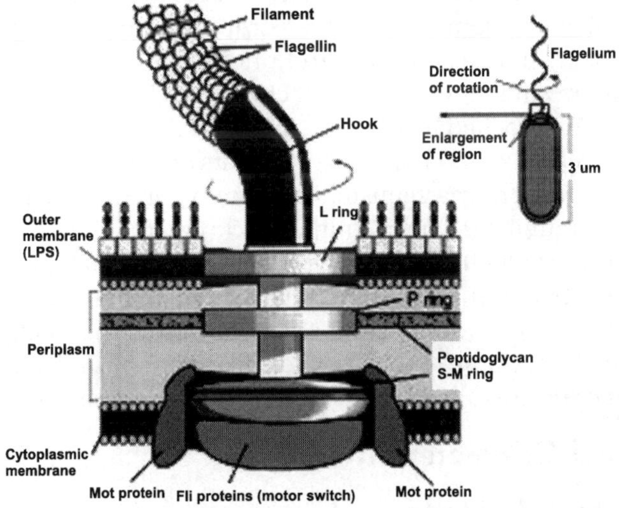

Figure 7.9: Structure of Bacterial Flagella (Gram negative).

The hook and basal body are quite different from the filament. Slightly wider than the filament, the hook is made of different protein subunits. The basal body is the most complex structure of the flagellum. In E.coli and Gram negative bacteria, the body has four rings connected to central rod. The outer

L and P rings associate with the lipopolysaccharide and peptidoglycan layers. The inner M ring connects the plasma membrane. Gram positive have only twp basal body rings, an inner ring connected to the plasma membrane and an outer one probably attached to the peptidoglycan. The synthesis of flagella is a complex process involving atleast 20 to 30 genes. Flagellin subunits are transported through the filament's hollow internal core. When they reach the tip, the subunits spontaneously aggregate under the direction of a special filament cap so that the filament grows at its tip rather than at the base. Filament synthesis is an excellent example of selfassembly.

Flagellar Movement

The mechanism of flagellar movement in prokaryotes is different from eukaryotic flagella. The bacterium moves when the helix rotates as the filament is in the shape of rigid helix. The flagella act just like propellers on a boat. The direction of flagellar rotation determines the nature of bacterial movement. The movement in monotrichous bacteria stop and tumble randomly by reversing the flagellar rotation. The polar flagella, rotate counter clockwise during normal forward movement, whereas the cell itself rotates slowly clockwise. Peritrichous bacteria also operate in a similar way. To move forward, the flagella rotate counter clockwise. As they do so, they bend at their hooks to for a rotating bundle that propels them forward. Clockwise rotation of the flagella disrupts the bundle and the cell tumbles. Motility enables the bacterium to move toward a favorable environment or away from a particular stimulus called taxis. Chemotaxis (include chemicals) and phototaxis (include light). Bacteria do not always swim aimlessly but are attracted by such nutrients as sugars and amino acids, and are repelled by many harmful substances and bacterial waste products. Movement toward chemical attractants and away from repellents is known as chemotaxis. The mechanism ofchemotaxis in E.coli has been studied most. Forward swimming is due to counterclockwise rotation of the flagellum, whereas tumbling results from clockwise rotation. The bacteria must be able to avoid toxic substances and collect in nutrient-rich regions and at the proper oxygen levels. E.coli has four different chemoreceptors that recognize serine, aspartate and maltose, ribose and galactose and dipeptides respectively. These chemoreceptors often are called methyl-accepting chemotaxis proteins (MCPs).

Morphological Characteristics

Size and Shape of Bacteria

Because of their small µ Prokaryotes are among the smallest of all organisms (0.5 to 2.0 size, bacteria have a large surface-to-volume ratio. The smallest member of the genus is about 0.3µm in diameter. Even smaller cells have been reported like the nanobacteria or ultramicrobacteria appear to range from around 0.2µm to not less than 0.05µm.E. coli, a bacillusof about average size is 1.1 to µm wide by 2.0 to 6.0 µm long. Spirochaetes occasionally reach 500 µm

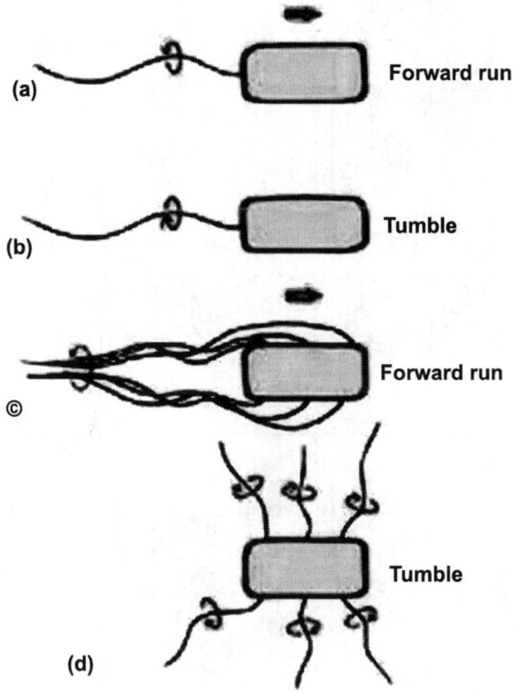

Figure 7.10: Bacterial Flagellar Movement.

in length and the Cyanobacterium oscillatoria is about 7 µm in diameter. The bacterium, Epulosiscium fishelsoni, can m in diameter). Thus a few bacteria are much larger µm long by 80 µbe seen with the naked eye (600 than the average eukaryotic cell (typical plant and animal cells are around 10 to 50 µm in diameter).

Shape and Arrangement

Typically bacteria have three basic shapes – spherical, rod like and spiral.

Spherical Bacteria

Coccus (pluralcocci)in pairs– diplococcic; in chains– Streptococci (Streptococcus); in cube like groups of eight– Sarcinae (Sarcina); in grape like structures– Staphylococci (Staphylococcus).

Rod like Bacteria

Bacillus (pluralbacilli) in pairs –diplococci; in chains – streptococci. Still others look like cocci and are called coccobacilli.

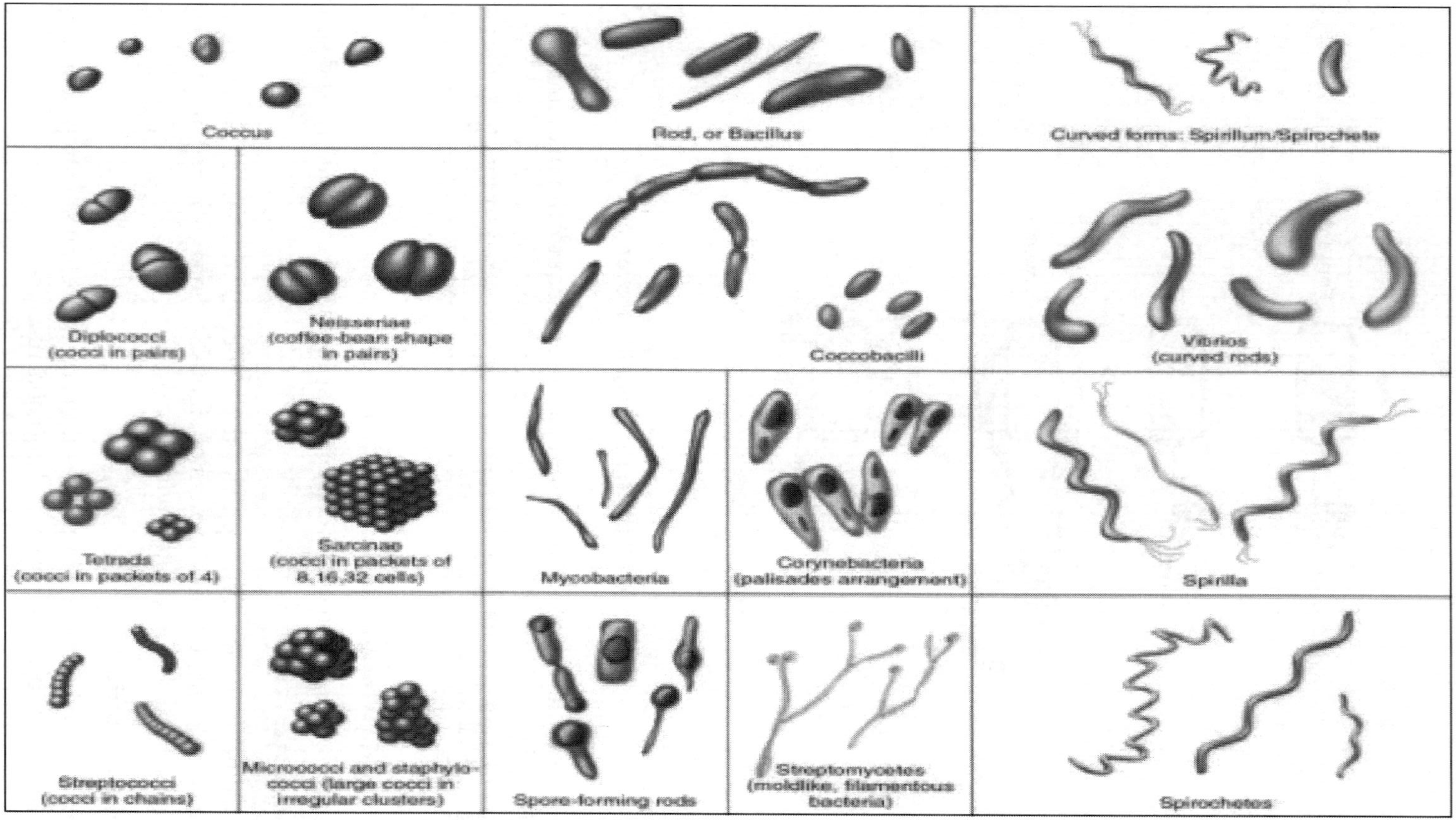

Figure 7.11: Different Morphological Characteristics and Diversity of Bacteria.

Spiral Bacteria

Classification of phytopathogenic prokaryotes, according to Bergey's manual of Systematic Bacteriology

A. Phenotypic classification
- I. Lmorphological
- II. Anatomical
- III. Staining
- IV. Cultural characteristics
- V. Nutrition
- VI. Environmental factors
- VII. Biochemical reactions
- VIII. Antigenic structure

B. Genotypic classification
- I. DNA-DNA hybridization
- II. G+C content

A.I. Morphological:
Based on shape and size
- Rod shaped (Bacillus)
- Spherical(Coccus)
- Spiral(Spirillum)
- Curved (Vibrio)

II. Anatomical:
Based on Flagellation
-FLAGELLATE
- Monotrichous
- Lophotrichous
- Amphitrichous
- Peritrichous

-AFLAGELLATE

Spore
- Spore former
- Non spore former

III.Staining:
-GRAM STAINING
- Gram positive cocci
- Gram negative cocci
- Gram positive rod
- Gram negative rod

- ACID FAST STAINING
- Acid fast bacilli
- Non acid fast bacilli

IV. Cultural characteristics
-EXTRA GROWTH FACTOR REQUIREMENT
- Fastidious
- Non fastidious

-UTILIZATON OF CARBOHYDRATE
- Oxidative
- Fermentative

-GROWTH RATE
- Rapid growers
- Slow growers

-PIGMENT PRODUCTION
- Pigment producer
- Pigment non producer

V. Based on nutrition:
- Autotroph
- Heterotroph

VI. Environmental factors
-TEMP
- Psychrophiles (15-20°c)
- Mesophiles (20-40°c)
- Extreme thermophiles (As high as 250°c)

-O: DEPENDANCE
- Acrobe
- Obligate Acrobe
- Microacropilic
- Facultative Anacrobe
- Obligate Anaerobe
- Capnophilic
- Aerotolerant

-pH
- Acidophiles
- Alkaliphiles
- Neutralophiles

- Salt conc.
  - Halophiles
  - Non halophiles
- Atm. pressure

Genetic Characteristics

- ☆ DNA Base composition: DNA molecules made up of base pairs guanine-cytosine and adenine thymine-base of classification-
- ☆ The sequence of nucleotides base in the sequence of nucleotide is unique for each type of organism most fundamental of all
- ☆ Pathogenicity- Ability to cause disease-Pathogenic to plants and to microorganism.
- ☆ Ecological- Marine microorganisms different from fresh water microorganism

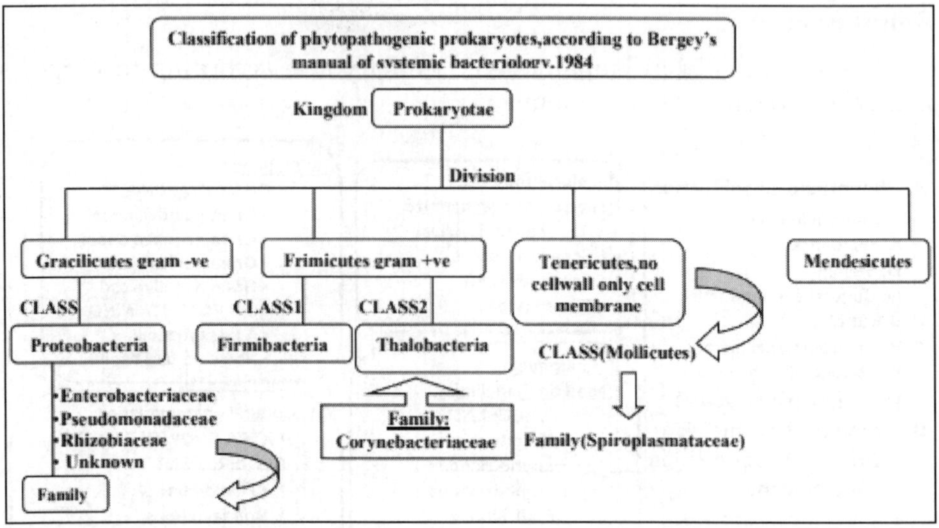

Reproduction of Bacteria

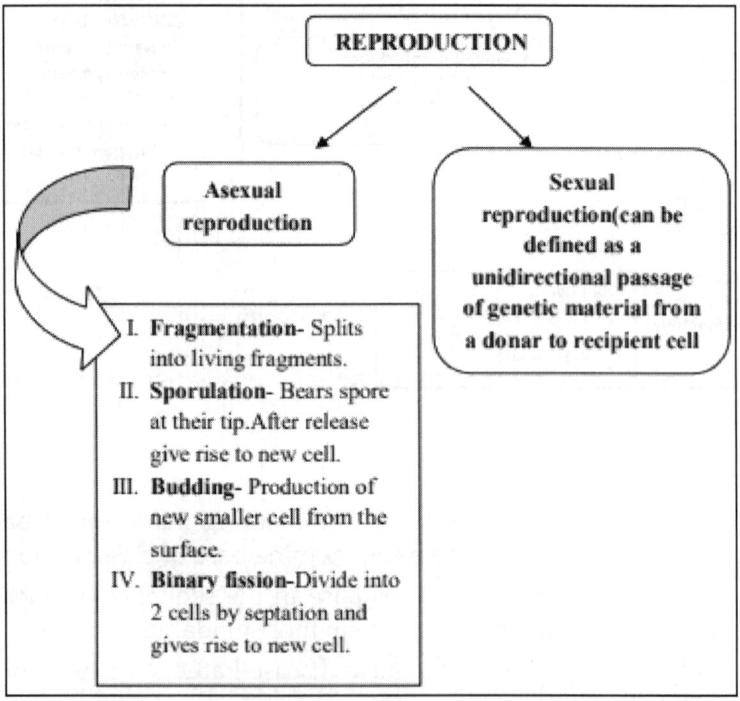

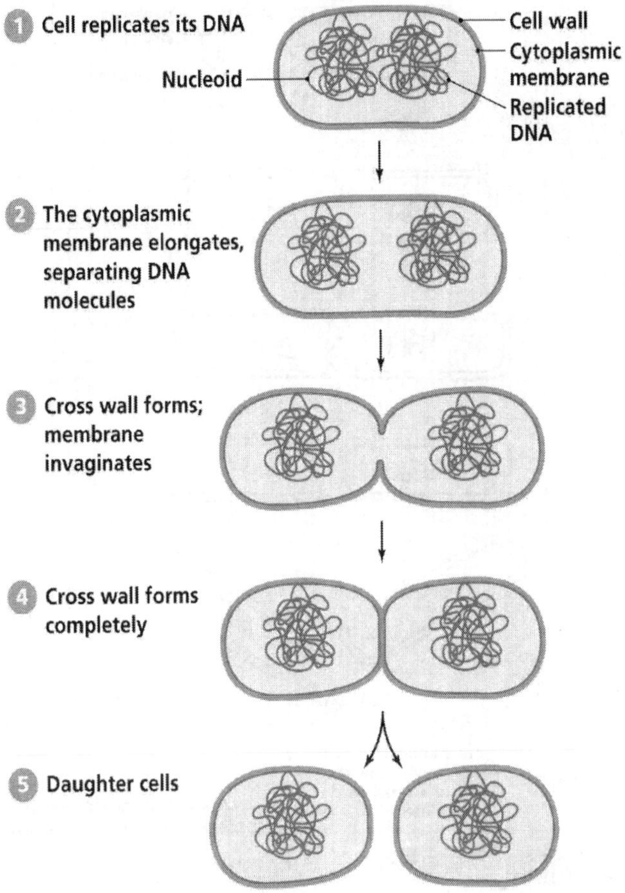
1 Cell replicates its DNA
Nucleoid
Cell wall
Cytoplasmic membrane
Replicated DNA
2 The cytoplasmic membrane elongates, separating DNA molecules
3 Cross wall forms; membrane invaginates
4 Cross wall forms completely
5 Daughter cells

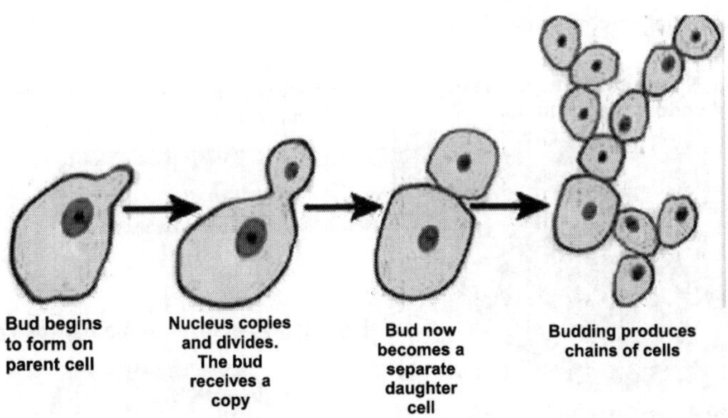
Bud begins to form on parent cell
Nucleus copies and divides. The bud receives a copy
Bud now becomes a separate daughter cell
Budding produces chains of cells

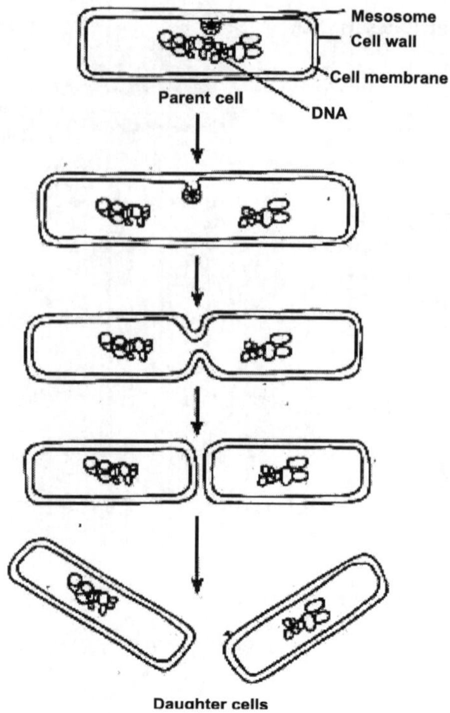
Mesosome
Cell wall
Cell membrane
Parent cell
DNA
Daughter cells

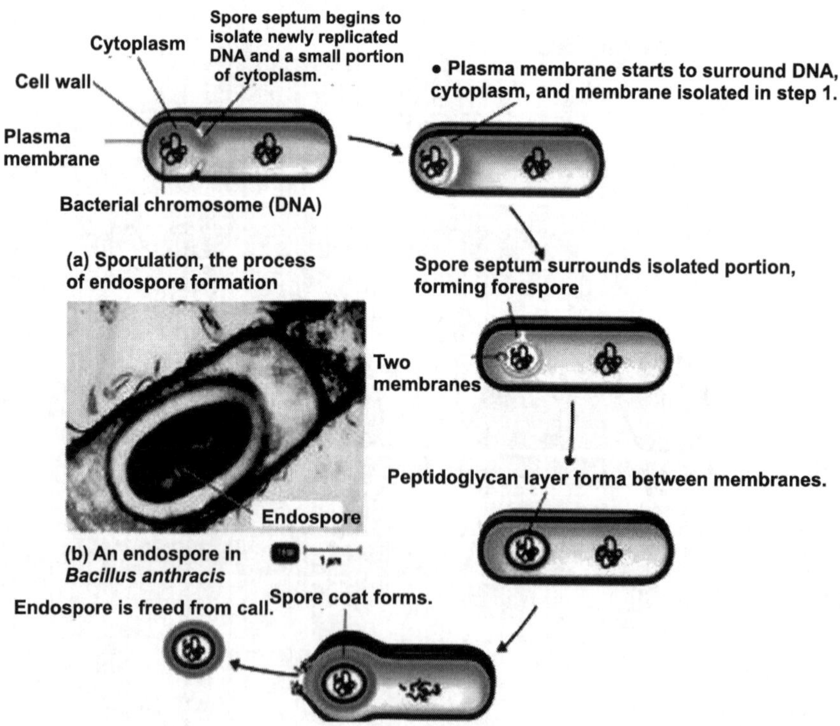
Cytoplasm
Cell wall
Plasma membrane
Spore septum begins to isolate newly replicated DNA and a small portion of cytoplasm.
Bacterial chromosome (DNA)
• Plasma membrane starts to surround DNA, cytoplasm, and membrane isolated in step 1.
(a) Sporulation, the process of endospore formation
Spore septum surrounds isolated portion, forming forespore
Two membranes
Peptidoglycan layer forma between membranes.
Endospore
(b) An endospore in *Bacillus anthracis*
1 µm
Endospore is freed from call.
Spore coat forms.

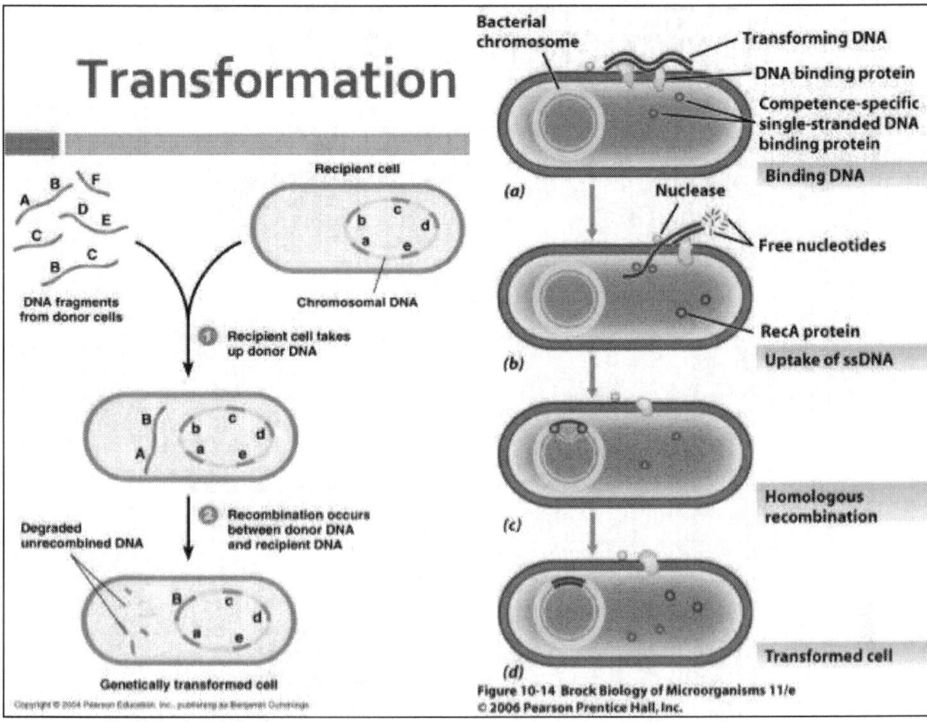
Transformation
A
B
F
D
E
C
B
C
DNA fragments from donor cells
Recipient cell
b
c
d
a
e
Chromosomal DNA
1 Recipient cell takes up donor DNA
B
A
b
c
d
a
e
2 Recombination occurs between donor DNA and recipient DNA
Degraded unrecombined DNA
B
c
d
a
e
Genetically transformed cell
Bacterial chromosome
Transforming DNA
DNA binding protein
Competence-specific single-stranded DNA binding protein
Binding DNA
(a)
Nuclease
Free nucleotides
RecA protein
Uptake of ssDNA
(b)
Homologous recombination
(c)
Transformed cell
(d)
Figure 10-14 Brock Biology of Microorganisms 11/e
© 2006 Pearson Prentice Hall, Inc.

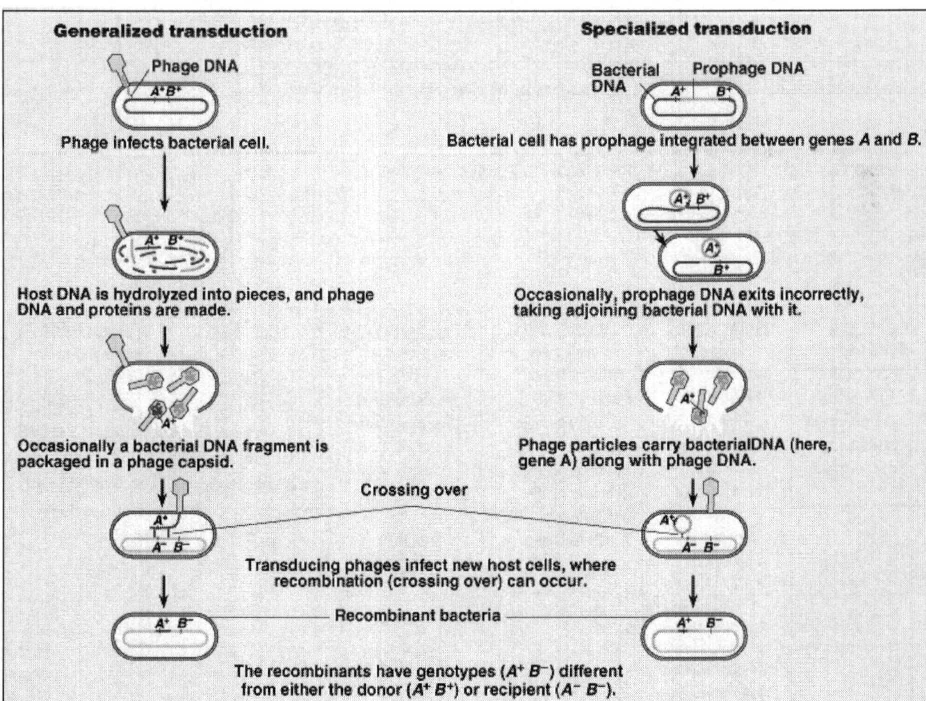
Generalized transduction
Phage DNA
A+B+
Phage infects bacterial cell.
A+ B+
Host DNA is hydrolyzed into pieces, and phage DNA and proteins are made.
A+
Occasionally a bacterial DNA fragment is packaged in a phage capsid.
A+
A– B–
Crossing over
Specialized transduction
Bacterial DNA
Prophage DNA
A+
B+
Bacterial cell has prophage integrated between genes A and B.
A+ B+
A+
B+
Occasionally, prophage DNA exits incorrectly, taking adjoining bacterial DNA with it.
A+
Phage particles carry bacterialDNA (here, gene A) along with phage DNA.
A+
A– B–
Transducing phages infect new host cells, where recombination (crossing over) can occur.
A+ B–
Recombinant bacteria
A+ B–
The recombinants have genotypes (A+ B–) different from either the donor (A+ B+) or recipient (A– B–).

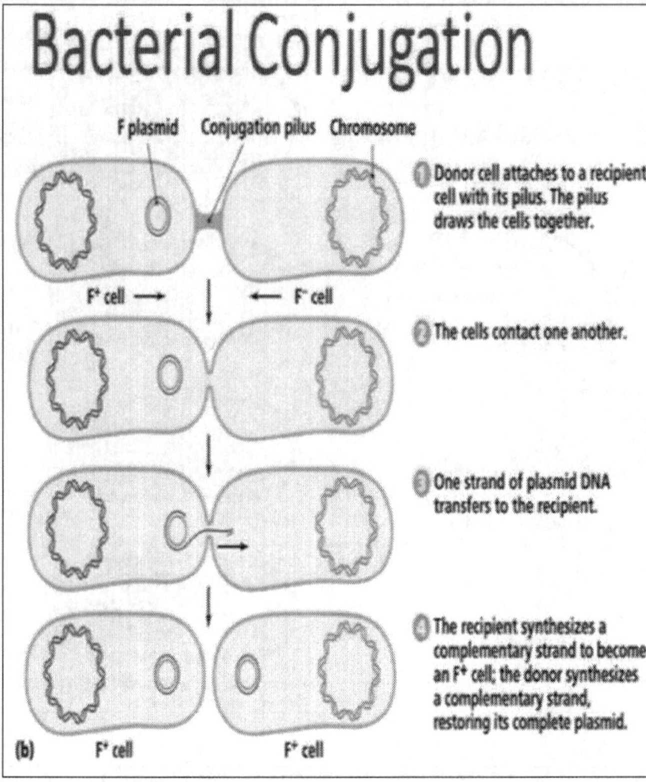

Cultivation of Bacteria

- **Autotroph** are those bacteria which utilize CO_2 as the major source of carbon. **Eg: Nitrosomonas**
- **Heterotroph** are those bacteria which use one/more org. compounds(glucose,amino acid)as their carbon source,they can't utilize CO_2. **Eg. Azospirilum, Staphylococcus,Lactobacillus**
- **Mesophyllus** can grow best within a temp. range of 25-40°c.All bacteria that are pathogenic are growing at this temp range. **Eg:Pseudomonas**
- **Thermophyllus** can grow best above 45°c.They are facultative thermophyllic (can grow in the range of mesophyllic).Eg **Bacillus thuringenesis**
- **Aerobic** requires O_2 for their growth and can grow when incubated in an air atmosphere
- **Anaerobic** they don't require O_2 for their growth and can't grow when incubated in air atmosphere

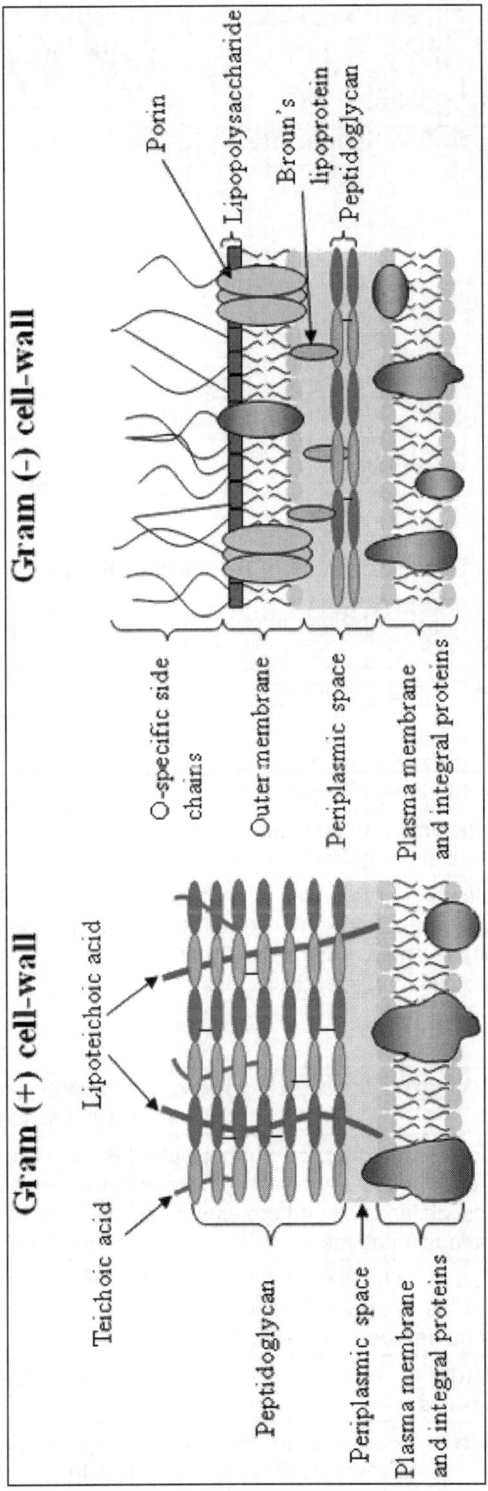
Gram (+) cell-wall
Teichoic acid
Lipoteichoic acid
Peptidoglycan
Periplasmic space
Plasma membrane and integral proteins
Gram (-) cell-wall
O-specific side chains
Outer membrane
Periplasmic space
Plasma membrane and integral proteins
Porin
Lipopolysaccharide
Broun's lipoprotein
Peptidoglycan

Comparative Characteristics of Gram-Positive and Gram-Negative Bacteria 12

Characteristic	Gram-positive	Gram-negative
Gram reaction	Retain crystal violet dye and stain dark violet or purple	Can be decolorized to accept counterstain (safranin); stain red
Peptidoglycan layer	Thick (multilayered)	Thin (single-layered)
Toxins produced	Primarily exotoxins	Primarily endotoxins
Resistance to drying	High	Low
Outer membrane	Absent	Present
Lipid and lipoprotein content	Low	High
Lipopolysaccharide (LPS) conten	Virtually none	High

Characteristics of Bacteria Cell Structures

Structure	*Functions(s)*	*Predominant Chemical composition*
Flagella	Swimming movement	Protein
Pili		
Sex pilus	Stabilizes mating bacteria during DNA transfer by conjugation	Protein
Common pili or fimbriae	Attachment to surfaces; protection against phagotrophicengulfment	Protein
Capsules (includes "slime layers" and glycocalyx)	Attachment to surfaces; protection against phagocytic engulfment, occasionally killing or digestion; protectionagainst desiccation	Usually polysaccharide; occasionally polypeptide
Cell wall		
Gram-positive bacteria	Confers rigidity and shape on cells	Peptidoglycan (murein) complexed with teichoic acids
Gram-negative bacteria	Confers rigidity and shape; outer membrane is permeability barrier; associated LPS and proteins have various functions	Peptidoglycan (murein) surrounded by phospholipid protein- lipopolysaccharide"outer membrane"
Plasma membrane	Permeability barrier; transport of solutes; energy generation; location of numerous enzyme systems	Phospholipid and protein
Ribosomes	Sites of translation (protein synthesis)	RNA and protein
Inclusions	Often reserves of nutrients; additional specialized functions	Highly variable; carbohydrate,lipid, protein or inorganic

Structure	Functions(s)	Predominant Chemical composition
Chromosome	Genetic material of cell	DNA
Plasmid	Extra chromosomal genetic material	DNA

Mollicutes

Mollicutes is a class of bacteria that are characterized by their lack of a cell wall. The name "Mollicutes" is derived from the Latin words "mollis," meaning soft, and "cutis," meaning skin, reflecting the soft and flexible nature of their cell membranes. The absence of a cell wall makes them unique among bacteria and gives them certain structural and functional characteristics.

One of the most well-known genera within the Mollicutes class is Mycoplasma. Mycoplasma species are often associated with various infections in humans, animals, and plants. They are known for their small size, ranging from 0.2 to 0.3 micrometers, and their ability to cause a variety of diseases.

Because of their lack of a cell wall, Mollicutes are resistant to antibiotics that target bacterial cell wall synthesis, such as penicillin. Instead, they are often treated with antibiotics that affect other aspects of bacterial metabolism.

Mollicutes play a significant role in both pathogenic and symbiotic relationships. Some species can cause diseases in humans and animals, while others form symbiotic associations with plants and insects. The study of Mollicutes is important in understanding bacterial diversity, evolution, and their impact on various ecosystems and hosts.

Subcategory: Mycoplasmataceae

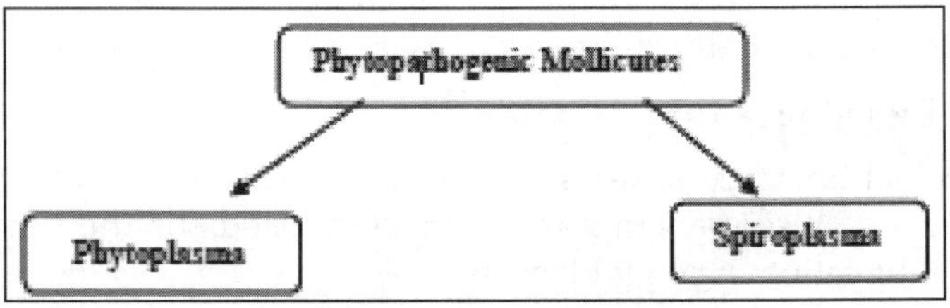

Phytoplasma

Phytoplasmas are specialized bacteria that belong to the class Mollicutes. Like other members of the Mollicutes class, phytoplasmas lack a cell wall, making them different from typical bacteria. Instead, they have a flexible cell membrane surrounded by a trilayered unit membrane. Phytoplasmas are primarily known for their association with plant diseases.

Characteristics of Phytoplasmas

Cell Structure

- Phytoplasmas have a unique helical or filamentous shape, and they lack a rigid cell wall.
- The absence of a cell wall makes them pleomorphic, meaning they can take on different shapes.

Plant Pathogens

- Phytoplasmas are obligate plant pathogens, meaning they rely on plants as their primary host.
- They cause diseases in a wide variety of plants, including crops, ornamental plants, and trees.

Transmission

- Phytoplasmas are typically transmitted between plants by insect vectors, such as leafhoppers and planthoppers.
- When an infected insect feeds on a plant, the phytoplasma can be transmitted to the plant.

Symptoms

- Phytoplasma infections can lead to a range of symptoms in plants, including yellowing of leaves, stunting, witches' broom (excessive branching), phyllody (abnormal development of floral parts into leaf-like structures), and overall decline in plant health.

Economic Impact

- Phytoplasma diseases can have significant economic consequences, affecting agricultural productivity and causing losses in crop yields.

Molecular Techniques for Detection

- Molecular techniques such as polymerase chain reaction (PCR) and DNA sequencing are commonly used for the detection and identification of phytoplasmas.
- Examples of diseases caused by phytoplasmas include "aster yellows," which affects a wide range of plants, and "stolbur," which affects tomatoes and other crops. Managing phytoplasmal diseases often involves controlling the insect vectors, using disease-resistant plant varieties, and implementing proper sanitation practices.
- Understanding phytoplasmas and their interactions with plants is crucial for developing effective strategies to manage and control the diseases they cause in agricultural and horticultural settings.

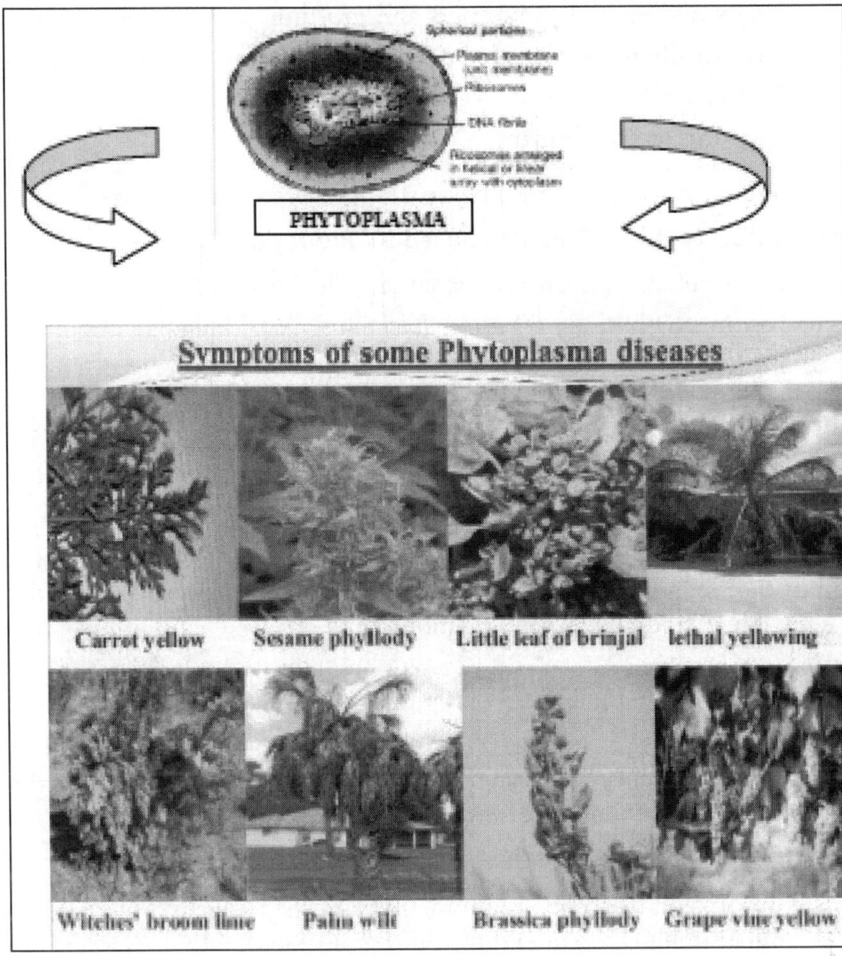

Characters of Phytoplasma

- ☆ Size vary from 100 nm-1µm in diameter.obligate parasite.
- ☆ Wall less,covered by trilamellar nit membrane with lipoproteins.
- ☆ Cells pleomorphic
- ☆ Aerobic to facultative anaerobes,nonflagellate,non-spore-formers and gram -ve
- ☆ Found intercellularly in phloem vessels in plant.
- ☆ Ribosomes of bacteria type.
- ☆ Genome is made up of double standed DNA.
- ☆ Osmotically stable.
- ☆ Self-reeplicable by transverse binary fission.
- ☆ Resistant to penicillin but sensitive to tetracycline and chloramphenicol.

- ☆ Transmission by leaf hoppers and grafting.
- ☆ Genetically more relates to acholeplasms than mycoplasmas.
- ☆ Can be controlled by cross-protection.
- ☆ Requires sterol for growth.

Can be controlled by thermotherapy by growing plants at 34-40°c

e.g. **Mulberry dwarf Disease caused by Phytoplasma**

- ☆ Aster yellows (*Macrosteles fascifrons*) (leaf hopper).
- ☆ Sesamum phyllody (*Orosius albicinctus, O. orientalis*) (leaf hopper)
- ☆ Little leaf of brinjal (*Hishimonas phycitis*) (leaf hopper)
- ☆ Grassy shoot of sugarcane (*Aphis maydis*) aphid *etc.*

Spiroplasma

Spiroplasma is a genus of bacteria that belongs to the class Mollicutes, similar to phytoplasmas. Spiroplasmas are helical or corkscrew-shaped bacteria and, like other members of Mollicutes, they lack a rigid cell wall. Instead, they have a flexible, triple-layered cell membrane.

Characteristics of Spiroplasma

Cell Structure

- ☆ Spiroplasmas are characterized by their helical or spiral-shaped cells, which are flexible and lack a peptidoglycan cell wall.

Host Range

- ☆ Spiroplasmas infect a wide range of hosts, including plants, insects, and animals. They are known to be associated with diseases in insects and plants.

Transmission

- ☆ In insects, Spiroplasmas are often transmitted by insect vectors, such as leafhoppers and planthoppers. They can also be transmitted vertically from infected females to their offspring.

Pathogenicity

- ☆ Spiroplasmas can be pathogenic and cause diseases in their host organisms. In insects, they are associated with various symptoms, including deformities, reduced fertility, and mortality.

Plant Pathogens

- ☆ Some Spiroplasma species are known to be plant pathogens, causing diseases in crops. For example, Spiroplasma citri is associated with citrus stubborn disease, which affects citrus plants.

Molecular Techniques for Detection

- ☆ Molecular techniques, such as PCR (Polymerase Chain Reaction) and DNA sequencing, are commonly used for the detection and identification of Spiroplasmas.

Symbiotic Relationships

- ☆ While some Spiroplasma species are pathogenic, others have been found to establish mutualistic or symbiotic relationships with their hosts. In some cases, they may provide benefits to their host organisms, such as protection against parasitoid wasps in insects.
- ☆ The study of Spiroplasma is important for understanding the diversity of bacteria, their ecological roles, and their impact on various host organisms. Additionally, research on Spiroplasmas contributes to the development of strategies for managing diseases they cause in agriculture and insect populations.
- ☆ Helical,wall less prokaryotes,phloem of diseased plants,helical in culture, a kind of mycoplasma,but can't be cultured.

Important Characteristics

- ☆ Helical in liquid media,gram+ve
- ☆ Size vary from 100-240nm in diameter,no flagella.
- ☆ Grown on culture medium in laboratory,colony appear like "fried egg" in culture medium.
- ☆ Multiplication by transverse binary fission.
- ☆ Resistant to penicillin but sensitive to tetracycline.

Important Disease and Vectors

Leaf Hopper

Leafhoppers are insects that belong to the order Hemiptera and are known for their ability to transmit various plant pathogens, including bacteria, viruses, and phytoplasmas.

Some important diseases and their associated vectors, particularly leafhoppers:

Pierce's Disease of Grapes

- ☆ **Pathogen:** *Xylella fastidiosa*, a bacterium.
- ☆ **Vector:** Glassy-winged sharpshooter (Homalodisca vitripennis) is a leafhopper species known to transmit Xylella fastidiosa, causing Pierce's disease. This disease affects grapevines, leading to symptoms such as leaf scorching and vine decline.

Maize Bushy Stunt Disease

- **Pathogen:** Spiroplasma kunkelii, a spiroplasma bacterium.
- **Vector:** The maize leafhopper (*Dalbulus maidis*) is the primary vector of Spiroplasma kunkelii, causing maize bushy stunt disease. This disease affects corn (maize) and can result in stunting, bushy growth, and reduced yield.

Citrus Stubborn Disease

- **Pathogen:** Spiroplasma citri, a spiroplasma bacterium.
- **Vector:** Several leafhopper species, including the beet leafhopper (*Circulifer tenellus*), can transmit Spiroplasma citri. This disease affects citrus trees, leading to stunted growth and misshapen fruit.

Rice Hoja Blanca

- **Pathogen:** Rice hoja blanca virus (RHBV).
- **Vector:** The brown planthopper (*Nilaparvata lugens*) is a significant vector of RHBV, causing rice hoja blanca disease. This viral disease affects rice plants, resulting in leaf yellowing, stunting, and reduced grain yield.

Tomato Big Bud Disease

- **Pathogen:** Candidatus *Phytoplasma solani*, a phytoplasma.
- **Vector:** The beet leafhopper (*Circulifer* tenellus) is implicated in the transmission of Candidatus *Phytoplasma solani*, causing tomato big bud disease. This disease affects tomatoes, leading to the development of large, distorted buds.

Potato Purple Top Phytoplasma

- **Pathogen:** Phytoplasma species causing potato purple top disease.
- **Vector:** Leafhoppers, including the beet leafhopper, can transmit phytoplasmas responsible for potato purple top disease. This disease affects potatoes, causing purple discoloration and stunting.

Control strategies for diseases transmitted by leafhoppers often involve managing the insect vectors through the use of insecticides, cultural practices, and the development of resistant plant varieties. Additionally, understanding the biology and ecology of both the leafhoppers and the associated pathogens is crucial for effective disease management in agricultural settings.

Chapter 8

Viruses and Plant Diseases

Introduction: An Overview

Viruses are microscopic infectious agents that can infect various living organisms, including plants. Plant viruses are responsible for a significant number of diseases in crops and ornamental plants, leading to economic losses in agriculture. Unlike animals, plants lack an immune system comparable to that of animals, making them more susceptible to viral infections.

Some Key Points about Viruses and Plant Diseases

Mode of Transmission

- **Vector-Borne:** Many plant viruses are transmitted by vectors, such as insects (aphids, leafhoppers, and whiteflies), nematodes, and fungi. These vectors act as carriers, moving the virus from infected plants to healthy ones.
- **Seed Transmission:** Some plant viruses can be transmitted through infected seeds, leading to the spread of the virus to new plants during germination.

Symptoms

- Plant viruses can cause a variety of symptoms, including yellowing of leaves (chlorosis), mosaic patterns, stunting, necrosis (cell death), and deformities in leaves and fruits.
- Symptoms may vary depending on the type of virus, the host plant, and environmental conditions.

Types of Plant Viruses

- **RNA and DNA Viruses:** Plant viruses can have either RNA (ribonucleic acid) or DNA (deoxyribonucleic acid) as their genetic material.
- **Single-Stranded or Double-Stranded:** The genetic material can be single-stranded or double-stranded, and it can be positive-sense or negative-sense.

Management and Control

- **Resistant Varieties:** Developing and planting virus-resistant crop varieties is one of the most effective strategies for managing viral diseases in plants.
- **Vector Control:** Managing and controlling the vectors that transmit plant viruses can help reduce the spread of the diseases.
- **Sanitation Practices:** Removing and destroying infected plants, practicing good hygiene in agricultural practices, and avoiding the use of contaminated tools can help prevent the spread of viruses.

Detection and Diagnosis

- **Serological Tests:** Enzyme-Linked Immunosorbent Assay (ELISA) and other serological tests can be used to detect the presence of specific plant viruses.
- **Molecular Techniques:** Polymerase Chain Reaction (PCR) and other molecular methods help in identifying and characterizing plant viruses at the genetic level.

Economic Impact

- Plant diseases caused by viruses can have a significant economic impact on agriculture, affecting crop yield and quality.
- Global trade can contribute to the spread of viral diseases, as infected plants or plant products may be transported across regions and countries.

Understanding the biology of plant viruses and implementing effective management strategies are crucial for sustainable agriculture and minimizing the impact of viral diseases on crops. Research and advancements in biotechnology also play a role in developing new tools for virus detection and crop protection.

History of Plant Viruses

The beginnings of plant virology date back to the late 19 th century, when Dutch microbiologist Martinus Beijerinck and Russian researcher Dmitrii Iwanowski were investigating the cause of a mysterious disease of tobacco.

These researchers independently described an unusual agent that caused mosaic disease in tobacco. What distinguished this agent from other disease-causing agents was its much smaller size compared to that of other microbes. This agent, later named *Tobacco mosaic virus* (TMV), was the first virus to be described.

Since then, a large number of diverse viruses have been found in plants, animals, fungi, and bacteria. The current estimate of recognized viruses is approaching 4,000, of which about 1,000 are plant viruses. The main reason that we study plant viruses is the negative impact that viral diseases have on crop production. Historically, viruses are perceived almost exclusively as a health threat to humans, livestock, and crop plants. However, recent progress in understanding virus-host interactions has transformed viruses into important tools of biomedicine and biotechnology.

For instance, plant viruses are being used to produce large quantities of proteins of interest in plants and to develop safe and inexpensive vaccines against human and animal viruses.

Basic Biology

Viruses represent not just another group of pathogens, but rather a fundamentally different form of life. Unlike all other living organisms, viruses are non-cellular. In contrast to cells, which multiply by dividing into daughter cells, viruses assemble from pools of their structural components. Mature virus particles are dormant; they come alive and reproduce only inside infected cells.

In other words, viruses are obligate parasites that cannot be cultivated using any growth media suitable for bacterial, fungal, plant or animal cell types. All viruses lack protein-synthesizing and energy-producing apparatuses. As a rule, virus particles are immobile outside the infected host; they rely on the aid of other organisms or the environment for their dissemination.

Morphology

There is a simple structural principle that applies to virtually all viruses in their mature form. Virus particles (virions) are composed of two principal parts, the genome that is made of nucleic acid, and a protective shell that is made of protein. In addition, some virus particles are enveloped by an outer membrane containing lipids and proteins (lipoprotein membrane).

The protein shells of plant viruses (capsids) are assembled in accord with one of the two fundamental types of symmetry. The first type of virion is helical (roughly elongated). The elongated viruses come in two major variants, rigid rods (Figure 8.1) and flexuous filaments (Figure 8.2). In both of these variants, the nucleic acid is highly ordered: it assumes the same helical conformation as the proteinaceous capsid.

The second type of virus particle is icosahedral (roughly spherical; Figure 8.3); the variations of this basic shape include bacilliform virions (Figure 8.4)

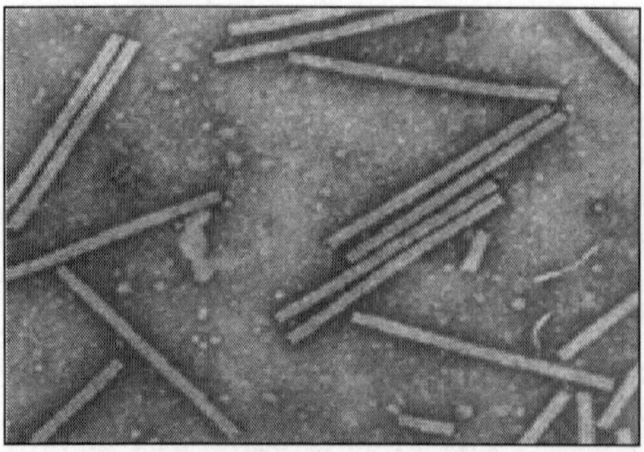

Figure 8.1

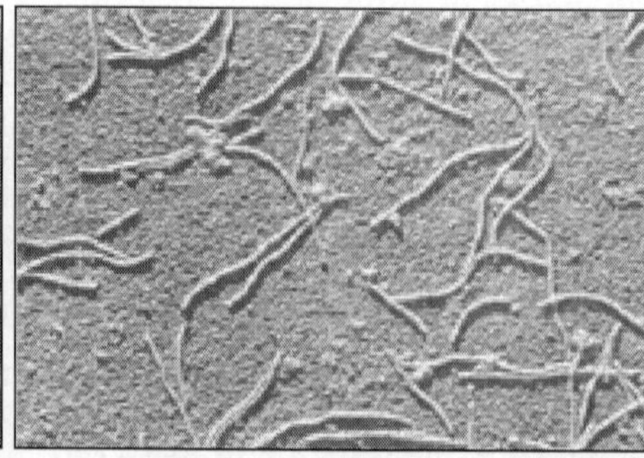

Figure 8.2

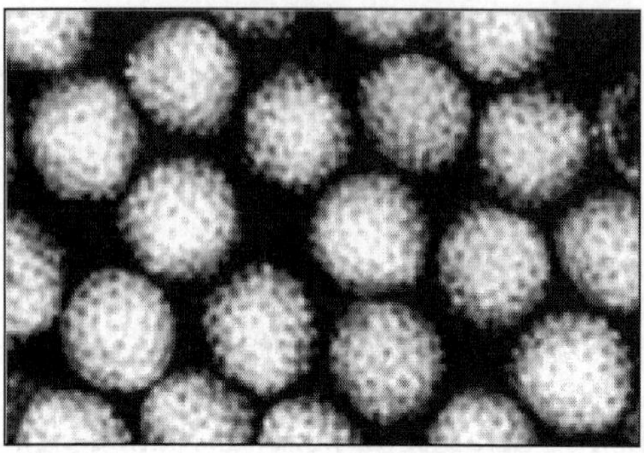

Figure 8.3

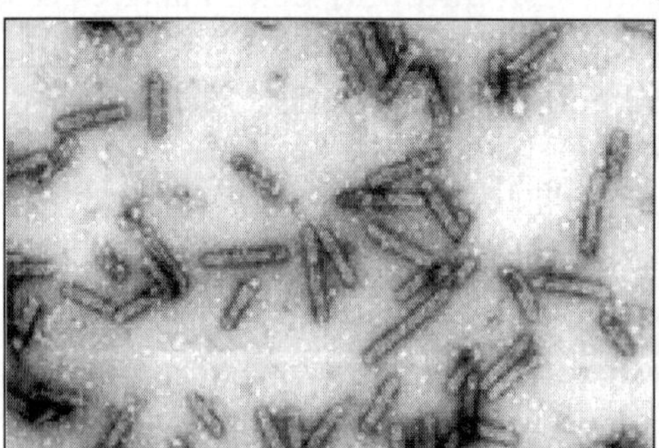

Figure 8.4

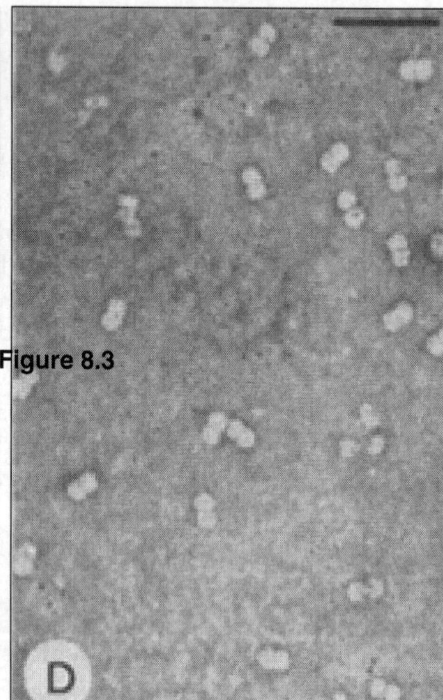

Figure 8.5

and twin virions composed of two joined incomplete icosahedra (Figure 8.5). In the icosahedral virions, the genomic nucleic acid forms a partially ordered ball inside the proteinaceous capsid. The icosahedral and elongated virions alike can self-assemble in a test tube if the nucleic acid and protein subunits are incubated under proper conditions.

Viruses are the smallest among all known organisms. The typical diameter of a spherical plant virus is ~30 nm. The rigid, rod-shaped TMV particle is 300 x 18 nm and consists of an RNA genome of about 6,400 nucleotides encapsidated by 2,130 copies of the TMV coat protein. Some of the filamentous viruses reach the length of ~2000 nm or 2 µm. For comparison, the typical size of a leaf mesophyll cell is ~50 µm.

Reproduction

As in other organisms, the information for virus reproduction is contained within its genome (Figure 8.6). Although the genetic material for most organisms is double-stranded (ds) DNA, only a minority of plant viruses possess dsDNA genomes. Some of the plant viruses have genomes that are composed of single-stranded (ss) DNA. However, the majority of plant viruses do not use DNA at all. Instead, the genomes of nearly all plant viruses are made of RNA. Most of these genomes are composed of ssRNA that is the same (positive-sense) polarity as the messenger RNAs of the cell. Some of the RNA viruses use ssRNAs of negative polarity, and yet others have genomes made of dsRNA. Due to this enormous variation in the very nature of the genetic material of viruses, the reproductive cycles and life styles of different viruses are often very distinct from each other.

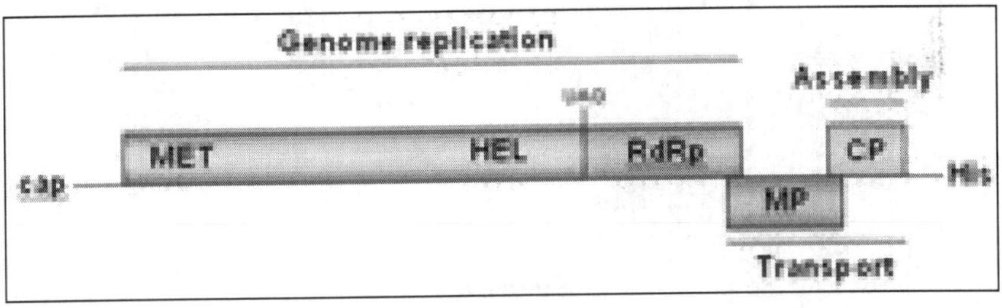

Figure 8.6

Since plant viruses are obligate, biotrophic parasites, their life cycles start by penetration of the virion into the cell. Plant viruses are unable to penetrate the plant cuticle and cell wall. It is believed that the virion enters the cytoplasm of the cell passively through wounds caused by mechanical damage to the cuticle and cell wall. The next phase of virus infection is the partial or complete removal of the coat protein shell of the virion in the cytoplasm. Next the cell mediates expression of the viral genome by providing a transcription apparatus (for DNA viruses) and a translation apparatus (for all viruses). The DNA viruses must be transported to the nucleus for transcription in order to gain access to the cell proteins required for the production of messenger RNA from viral DNA. Translation of viral RNA in the cytoplasm produces viral proteins that are required for completion of the virus life cycle.

All viruses must direct the formation of at least three types of proteins: replication proteins that are essential for nucleic acid production, structural proteins that form the protein shell and other components contained in the virions, and movement proteins that mediate virus transport between plant cells (Figure 8.6). The viral replication proteins combine with cellular proteins to produce a complex of proteins that manufactures multiple copies of the virus genome. These newly made genomes interact with the structural proteins to form new virions. The diagram below (Figure 8.7) illustrates the steps that occur within the cell during the replication of *Tobacco mosaic virus*.

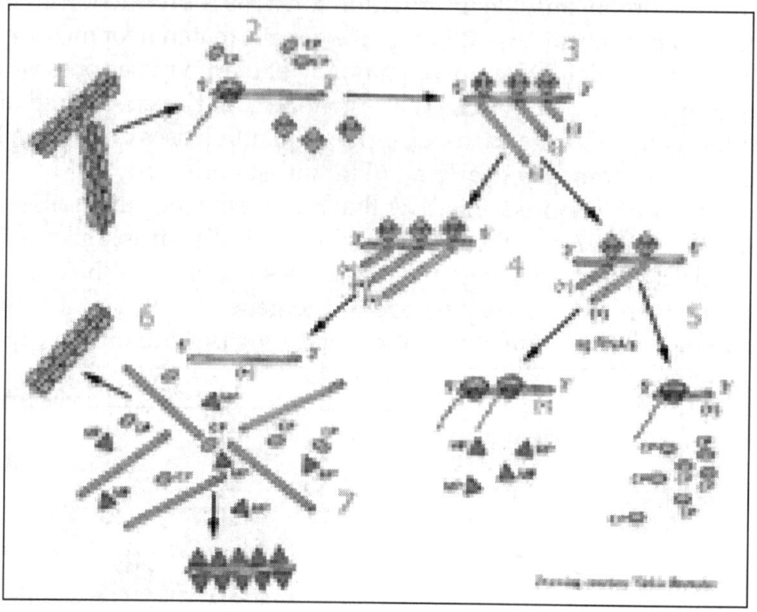

Figure 8.7

The next step in the virus reproduction cycle is movement of the virus into neighboring cells. Depending on the virus, the viral genomes or the virions are transported into neighboring cells through small channels called plasmodesmata that form connections between cells. Many viruses produce movement proteins that modify the plasmodesmata channels and facilitate viral movement into neighboring cells. The following diagram (Figure 8.8) illustrates the movement of *Tobacco mosaic virus* from an infected cell to a neighboring cell.

The process of cell-to-cell movement is relatively slow: it takes from one to a few hours for a virus to multiply in a cell and move to the next cell. To successfully colonize an entire plant, a virus needs to enter the vascular system of the plant. The process of systemic, or long-distance transport normally proceeds through the phloem sieve elements where viruses move passively with the flow of photosynthates. After quite rapid systemic spread of the virus (centimeters per hour) in the phloem, the virus moves from the phloem into surrounding cells where it reproduces and spreads by cell-to-cell movement. The time between initial infection of one or a few cells and systemic infection of the plant varies from a few days to a few weeks depending on the virus, host plant, and environmental conditions. Transmission of the virus from one plant to another (see section on survival and dissemination) completes the virus life cycle.

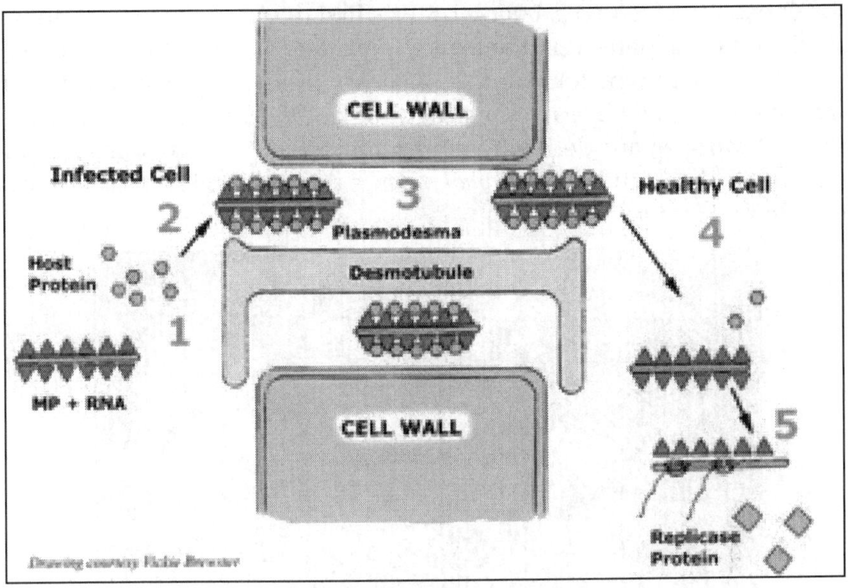

Figure 8.8

Systematics

Ideally, systematics (the study of the kinds of organisms and the relationships between them) should reflect the evolutionary history of a biological species. This history can be constructed by analysis of fossilized predecessors and by comparison of the genes of existing organisms (phylogenetic analysis). Since viruses left no fossils, their evolution can be reconstructed with some confidence only by phylogenetic analysis. This analysis works reasonably well at the intermediate taxonomic levels, such as virus genus and family. However, due to the nature of the viral form of life, it gets more complicated or even unfeasible at the lower (species) level or higher levels. Because the viruses exhibit very high rates of reproduction and mutation, each reproduction cycle results in a number of genetically similar, but not identical variants. Further selection and evolution of slightly different genomes results in a continuum of variants that range from point mutants to virus strains. Unlike cellular organisms, which are believed to originate from a common ancestor, the viruses most likely originated more than once. The diverse origins of different virus groups permit their evolutionary classification only within particular virus types, such as viruses that have positive-sense RNA genomes, or small ssDNA genomes.

When we consider plant virus classification, we should remember that it was the mid 1930s before the first plant virus was purified and characterized.

Prior to this time, most plant virologists named a virus based on the host plant in which it was found and the type of symptom that the virus caused in the plant. For example, *Tobacco mosaic virus* was first described in tobacco in which it induces a mosaic pattern in the leaves (Figure 8.9). Virus names are usually shortened to an acronym, for example, *Tobacco mosaic virus* is shortened to TMV, and *Tomato spotted wilt virus* is shortened to TSWV. The names of many virus genera and families are derived from an important virus within the family. For example, the family name *Bromoviridae* is derived from *Brome mosaic virus* which is in this family.

Figure 8.9

The largest subdivisions of plant viruses are defined by the chemical form of their genomes. By far, the largest number of plant viruses possess single-stranded (ss), positive-sense RNA genomes, and these viruses are called positive-strand RNA viruses. Examples of the most economically important families of the positive-strand RNA viruses are *Bromoviridae, Closteroviridae, Luteoviridae*, and *Potyviridae*. Relatively few plant viruses, exemplified by the families *Bunyaviridae* and *Rhabdoviridae*, possess negative-sense RNA genomes. *Reoviridae* is the largest family of the double-stranded (ds) RNA viruses.

There is only one family of the plant viruses with dsDNA genomes, the *Caulimoviridae* family or so-called pararetroviruses, and replication of these viruses involves an RNA intermediate. The true plant DNA viruses are represented by the large and economically important family *Geminiviridae*. These viruses possess ssDNA genomes, and they have a dsDNA intermediate in their life cycle. The evolutionary relationships among the positive-strand, negative-strand, and dsRNA viruses, as well as the pararetroviruses and ssDNA viruses appear to be extremely distant if not absent.

Survival and Dissemination

Only relatively few viruses, such as TMV, rely on long-term (up to decades) survival in the environment and on passive mechanical transmission from plant to plant (Ford and Evans 2003). Most plant viruses are actively transmitted from infected to healthy plants by a living organism called a vector. Plant-feeding arthropods, nematodes and plant-parasitic fungi are the major types of vector organisms for plant viruses. Among these, aphids and whiteflies (Figure 8.10) are capable of transmitting the largest number of virus species. Most viruses are actively transmitted to healthy plants in a matter of seconds, hours, or days by vectors.

Figure 8.10

In addition, some viruses are transmitted from infected plants when these plants are vegetatively propagated, for example, through tubers or grafting, or 'vertically' through seeds or pollen from infected plants. This type of transmission is particularly important for virus survival during the winter.

Transmission of viruses by a vector is a very specific process. Each particular virus can be transmitted by only one vector type (*e.g.*, aphid), and not by another (*e.g.*, whitefly). Conversely, each vector species (*e.g.*, green peach aphid *Myzus persicae*), can transmit some viruses (*e.g.*, *Beet yellows virus*) but not others (*e.g.*, *Citrus tristeza virus*), despite the fact that these viruses are genetically very similar to each other.

The interaction between a virus and its specific vector that results in virus transmission varies for different virus vectors. The relationships that have developed between viruses and their vectors are complex and of considerable interest to plant virologists because vectors provide the main method of spread for many viruses that cause severe economic losses.

For some virus/vector combinations, the interaction between a virus and its vector is very superficial and is the result of attachment to the external surfaces of the vector mouthparts. For example, viruses in the *Potyvirus* genus produce a special protein called helper component that "glues" the virions to aphid stylets. In this case, acquisition of the virus from infected plants and inoculation of virus to healthy plants takes from seconds to minutes. In contrast, viruses in the family *Luteoviridae* circulate in their aphid vector moving through the gut wall, into the body cavity, and then into the saliva of the aphid (Figure 8.11).

Figure 8.11

It takes at least 12 hours for these viruses to circulate in their aphid vector before they can be transmitted to another plant. For other plant virus vectors, the relationship is very intimate and the viruses actually multiply in the cells of their insect vectors.

For example, *Tomato spotted wilt virus* multiplies in cells of its thrips vector, and once thrips acquire this virus, they can transmit it for the rest of their life. These plant viruses that multiply in their insect vectors are considered to be both plant and insect viruses. Because management practices for plant diseases are often directed at the virus vectors, an understanding of the virus transmission process is critical for development of effective management strategies for diseases caused by plant viruses

Plant-Virus Interactions

Viruses are capable of infecting virtually all species of cultivated and wild plants. However, host ranges of individual viruses vary from very narrow to very broad. For instance *Citrus tristeza virus* infects only a few species in the *Citrus* genus, whereas *Cucumber mosaic virus* infects over 1000 species in 85 plant families. Susceptibility or resistance of plant species and cultivars to viruses is determined primarily by the plant genotype. Plants possess active and passive means of preventing virus infection.

Passive defenses are due to the failure of the plant to produce one or more host factors required for virus reproduction and spread within the host. Active defenses include detection and destruction of the virus-infected cells due to the function of specific resistance genes in the plant. Normally, resistance genes are active only against a particular virus. In addition, plants possess a general defense system that is somewhat analogous to the animal immune system. The major difference between the two is that the immune system in animals targets a pathogen's proteins, whereas the plant defense system, which is called RNA silencing, detects and degrades viral RNAs.

Depending on the particular combination of virus and host, and on environmental conditions, a plant's response to infection may range from a symptomless condition to severe disease and plant death. In some cases, small necrotic or chlorotic spots called local lesions (Figure 8.12) develop at the site of infection. In most cases, viruses spread throughout the whole plant and cause a systemic infection.

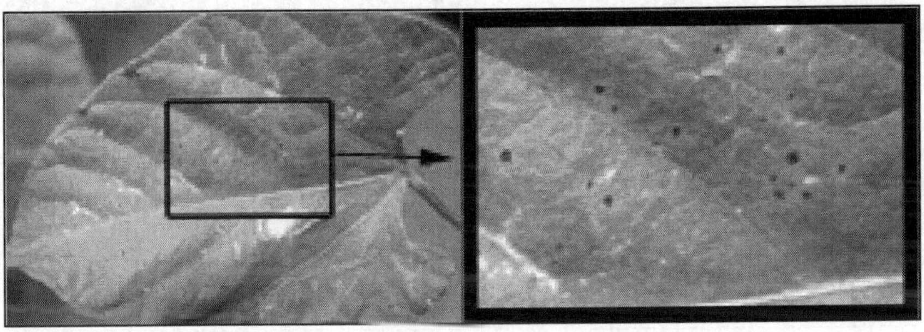

Figure 8.12

Typical leaf symptoms of viral diseases include mosaic patterns (Figure 8.9), chlorotic or necrotic lesions (Figure 8.13), yellowing, stripes or streaks (Figures 8.14, 8.15), vein clearing, vein banding (Figure 8.16), and leaf rolling and curling (Figure 8.17). Flower symptoms include deformation and changes in the color of the flowers including dramatic color mosaics called color breaking (Figures 8.18, 8.19). Fruit and vegetable symptoms may include mosaic patterns (Figure 8.20), stunting, discoloration or malformation (Figure 8.21), and chlorotic ringspots (Figures 8.22, 8.23). Stems of plants may develop stem pitting and grooving or tumors in response to virus infection (Figure 8.24).

The symptoms induced by plant viruses lead to reduced crop quality and yield. The extent of these crop losses is demonstrated by the following three examples. *Cacao swollen shoot virus* is estimated to cause an annual loss of 50,000 tons of cocoa beans in Africa.

Figure 8.13

Figure 8.14

Figure 8.15

Figure 8.16

Figure 8.17

Figure 8.18

Figure 8.19

Figure 8.20

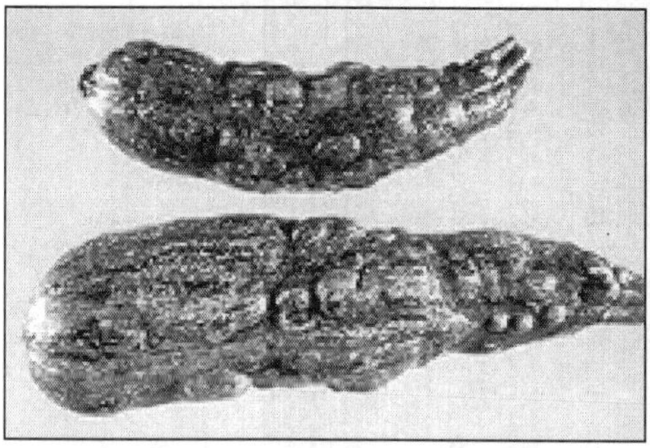

Figure 8.21

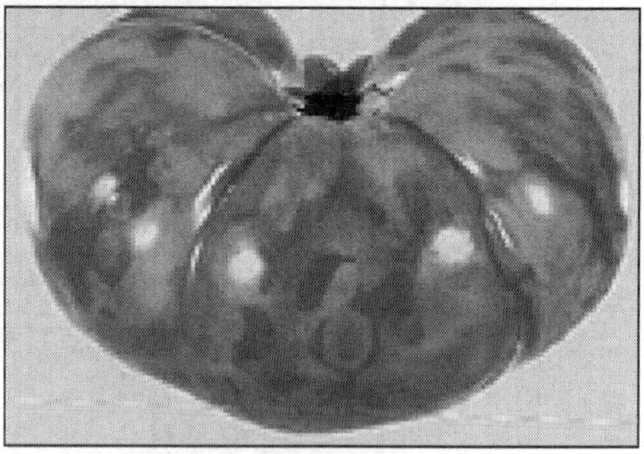

Figure 8.22

Figure 8.23

Figure 8.24

Diagnosis of Plant Virus Diseases

Because different viruses may elicit similar symptoms, the disease phenotype can provide only limited, although important, information for disease diagnosis. More specific and reliable methods of virus identification are based on various properties of the virus. These properties and corresponding approaches include the following:

- ✰ **Pathogenicity:** Bioassays using indicator plants. Some plant genera, such as *Nicotiana* (tobacco) and *Chenopodium* (lambsquarters) are hosts for a number of viruses. Since the responses of these plants to viral infections under greenhouse conditions are consistent and distinctive, they are commonly used as indicator plants (Walkey 1991). Two major types of responses are local lesions (Figure 8.12) which are confined to inoculated leaves (local lesion hosts), and systemic infections which produce symptoms on leaves distant from the inoculation site (systemic hosts). Many plant viruses are transmissible to indicator plants by means of mechanical transmission (Figure 8.25) or grafting.

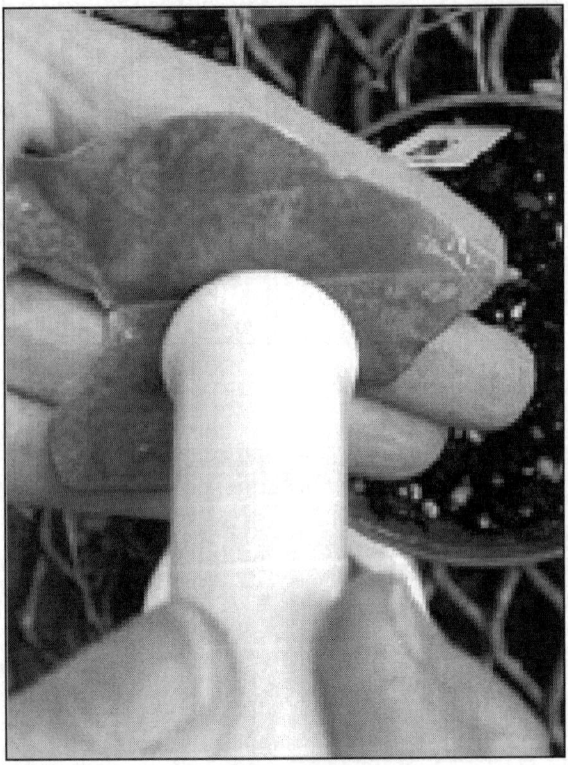

Figure 8.25

- ✰ **Transmissibility** : Vector transmission assays. Because of vector specificity, identification of the organism that transmits the virus may provide important information for virus identification.
- ✰ **Architecture of virus particles** – Electron microscopy. The shape and size of virions distinguish rod-shaped, filamentous, icosahedral, or

large enveloped particles (Figures 8.1-8.5). On the other hand, viruses sharing the same shape and size are difficult to distinguish by their appearance. For instance, small spherical viruses may be difficult to distinguish from each other and from plant ribosomes.

☆ **Presence of virus-specific structures in infected cells** – Electron microscopy. Because of their intimate association with components of the cell, viruses often form unusual structures within plant cells as a result of virus infection. For example, viruses in the family *Potyviridae* produce 'pinwheel' inclusions (Figure 8.26) that are characteristic of viruses in this family and are not found in healthy cells or cells infected with other viruses. Virus-specific inclusions have been characterized for a number of plant virus families and genera, and the detection of these inclusions indicates the presence of a virus within that group.

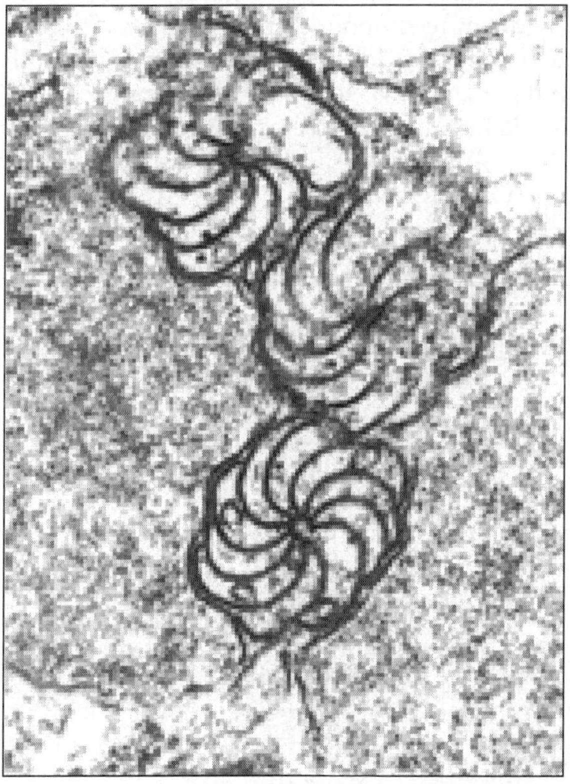

Figure 8.26

Properties of the protein coat – Immunological procedures. These tests rely on identification of a virus (the antigen) through its reaction with specific antibodies. Specific antibodies are produced by an animal when foreign proteins are introduced into the animal. To obtain antibodies that react specifically with a particular plant virus, scientists inject a purified preparation of the plant virus into an animal (usually a mouse or a rabbit). Several weeks after the injection, blood is collected from the animal and antibodies are separated from the blood. These antibodies react specifically to the viral proteins that were injected into the animal, and scientists have taken advantage of this specific interaction to develop laboratory tests for plant viruses.

One of the most widely used diagnostic tests for plant viruses is an antibody-based procedure called the Enzyme-Linked Immunosorbent Assay (ELISA). An example of one type of ELISA test is illustrated in Figure 8.27. In this test, an extract of plant tissue being tested for virus is incubated in the well of a plastic microtiter plate, and virions present in the extract bind to the surface of the well. Next, antibody to the virus is added to the well, and the antibody binds to the virus particles. A second antibody that has been conjugated, or linked, to an enzyme is added to the well where it attaches to the first antibody. Following a wash to remove unbound enzyme-antibody conjugate, a colorless substrate for the enzyme is added. The enzyme processes the substrate and produces a colored product (*e.g.*, yellow). The appearance of color indicates the presence of virus and the intensity of the color can be used to estimate the virus concentration. ELISA is by far the most widely used immunological technique for virus identification in agriculture.

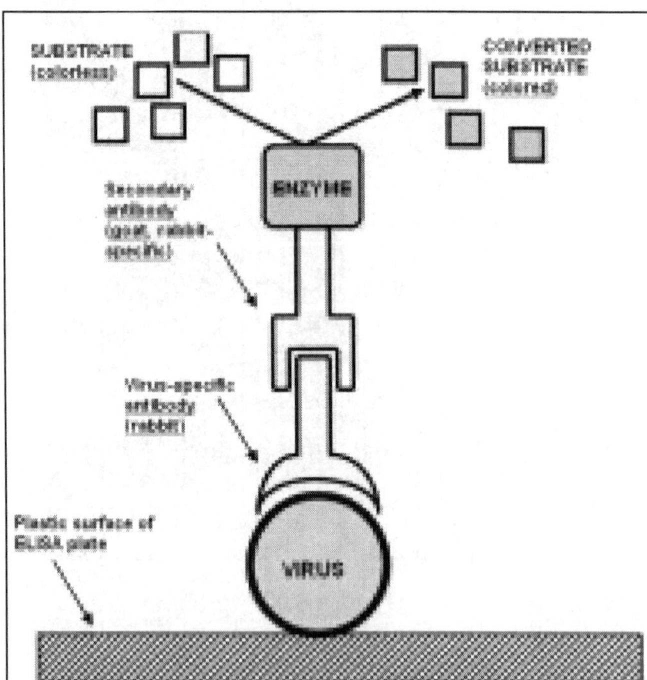

Figure 8.27

☆ **Properties of viral nucleic acid** – PCR amplification. Polymerase Chain Reaction (PCR) is a very sensitive and specific technique for virus detection that is based on the presence of unique nucleic acid sequences in the genome of a virus (Bartlett 2003). The viral DNA sequence that a researcher wants to amplify undergoes about thirty cycles of copying in a small test tube. An example of one type of PCR test is illustrated in Figure 8.28. In the PCR test, short pieces of DNA (primers) that are complementary to portions of the viral nucleic acid are hybridized (attached) to the viral nucleic acid. A heat-stable polymerase enzyme is then used to synthesize multiple copies of a strand of DNA (cDNA) that is complementary to the portion of the viral nucleic acid sequence between the two primers. During the PCR reaction, the number of molecules of the target sequence doubles with each cycle, and after 30 cycles over 10 billion copies of the target sequence are produced. The PCR product can be detected using gel electrophoresis, a process in which an electrical current is used to separate molecules in a gel based on their size. After separation, the DNA fragments in the gel are stained with a dye that is specific for nucleic acid, and the occurrence of a band of the appropriate size indicates the presence of the virus for which the DNA primers were designed. For RNA viruses, the

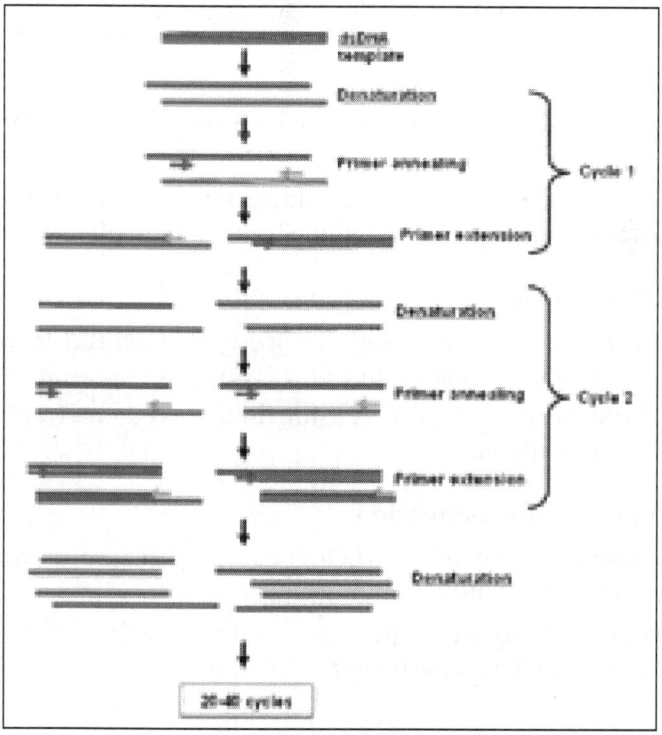

Figure 8.28

viral RNA strand is "reverse" transcribed into its DNA complement, followed by amplification of the resulting DNA using PCR.

Managing Plant Virus Diseases

Managing plant virus diseases involves a combination of preventive measures, cultural practices, and, in some cases, the use of chemical or biological control methods. Here are key strategies for the management of plant virus diseases:

Use of Resistant Varieties

- Planting resistant or tolerant varieties is one of the most effective and environmentally friendly approaches to managing viral diseases.
- Plant breeders work to develop cultivars that have genetic resistance to specific viruses.

Vector Control

- Managing the vectors that transmit plant viruses is crucial. This may involve the use of insecticides to control insect vectors like aphids, whiteflies, and leafhoppers.
- Cultural practices, such as removing weed hosts that can harbor both viruses and vectors, can help in vector control.

Sanitation Practices

- Remove and destroy infected plants promptly to prevent the further spread of viruses.
- Practice good sanitation in the field, greenhouse, or garden by cleaning tools and equipment to avoid the transmission of viruses.

Use of Certified Seeds

- Planting certified disease-free seeds is essential to prevent the introduction of viruses into a new crop.
- Seed certification programs ensure that the seeds are free from viruses and other pathogens.

Crop Rotation and Diversification

- Implementing crop rotation helps break the disease cycle by preventing the continuous cultivation of the same crop.
- Diversifying crops in a given area can reduce the buildup of specific viruses that affect a particular plant species.

Physical Barriers

- ☆ Use physical barriers such as insect-proof screens in greenhouses to prevent the entry of virus vectors.
- ☆ Install physical barriers between different crops to minimize the spread of viruses.

Early Detection and Diagnosis

- ☆ Regularly monitor crops for symptoms of viral diseases.
- ☆ Use diagnostic tools such as serological tests (ELISA) and molecular techniques (PCR) for accurate and early detection of viruses.

Biological Control

- ☆ Introduce natural enemies of virus vectors, such as predatory insects or parasitoid wasps, to control vector populations.
- ☆ Some beneficial microorganisms can also be used to inhibit the multiplication or movement of viruses within plants.

Chemical Control

In certain situations, chemical control measures may be employed to manage virus vectors. However, this approach is often considered a last resort due to potential negative impacts on beneficial organisms and the environment.

Education and Extension Services

- ☆ Educate farmers, growers, and agricultural workers about the symptoms of viral diseases and proper management practices.
- ☆ Extension services can play a crucial role in disseminating information about the latest advancements in virus management.

Global Collaboration

- ☆ International collaboration is important to prevent the global spread of plant viruses through trade and movement of plant material.
- ☆ Quarantine measures and adherence to international regulations can help control the movement of infected plants across borders.

It's important to note that the effectiveness of these management strategies may vary depending on the specific virus, the host plant, and local environmental conditions. Integrated pest management (IPM) approaches that combine multiple strategies often provide the most sustainable and effective control of plant virus diseases.

Viroids: Infectious Nucleic Acids

Viroids are small, circular, single-stranded RNA molecules that are infectious and capable of causing diseases in plants. Unlike viruses, viroids lack a protein coat (capsid) and do not encode any proteins. Their ability to infect plants and induce disease challenges the traditional concept of infectious agents requiring both nucleic acid and protein components.

In the intricate tapestry of microbial life, viroids emerge as remarkable entities, challenging conventional notions of infectious agents. Unlike traditional viruses that possess both nucleic acid and protein components, viroids are minimalist pathogens–small, circular, single-stranded RNA molecules devoid of a protective protein coat. Despite their seemingly simplistic structure, viroids wield a remarkable ability to infect plants and induce diseases. Their unique biology and pathogenicity have sparked scientific curiosity, prompting researchers to delve into the mysteries of these infectious nucleic acids. In this exploration, we unravel the defining features of viroids, their modes of transmission, the specific havoc they wreak on plant hosts, and the ongoing efforts to understand and manage the impact of viroid-induced diseases on agriculture. In the world of molecular parasites, viroids stand out as intriguing agents that challenge our understanding of infectious entities and their intricate dance with living organisms.

Characteristics of Viroids

Structure

- Viroids consist of a short, single-stranded RNA molecule, typically around 246 to 375 nucleotides in length.
- The RNA forms a closed, circular structure and does not code for any proteins.
- Viroids have a highly base-paired secondary structure.

Transmission

- Viroids are primarily transmitted through mechanical means, such as contaminated pruning tools or during agricultural practices that involve contact with infected plant material.
- Some viroids are also transmitted through pollen and seeds.

Host Range

- Viroids are known to infect plants, and each viroid species tends to have a specific host range, often limited to certain plant families.
- Common hosts include crops like tomatoes, potatoes, and citrus fruits.

Pathogenicity

- Viroids induce disease in plants by interfering with the host's normal cellular processes.
- Symptoms of viroid infection include stunted growth, leaf deformation, and changes in pigmentation.
- Viroids can also interfere with the host's ability to defend against other pathogens.

Replication

- Viroids replicate autonomously within plant cells.
- The replication process involves host enzymes, and viroids can utilize host machinery to replicate their RNA.

Examples of Viroids

- Potato Spindle Tuber Viroid (PSTVd): A viroid that infects potatoes and other solanaceous plants, causing tuber deformities.
- Citrus Exocortis Viroid (CEVd): A viroid that infects citrus trees, leading to symptoms such as bark scaling and stunting.
- Chrysanthemum stunt viroid (CSVd): A viroid affecting chrysanthemums, causing stunted growth and reduced flower size.

Detection

- Viroids can be detected using techniques such as molecular methods like reverse transcription-polymerase chain reaction (RT-PCR).
- Serological methods, including enzyme-linked immunosorbent assay (ELISA), are also employed for viroid detection.

Control and Management

- There are no specific antiviral drugs for treating viroid infections in plants.
- Management strategies include using disease-free planting material, practicing good sanitation to prevent mechanical transmission, and avoiding the use of contaminated tools.

Understanding viroids and their interactions with host plants is crucial for developing effective strategies to manage viroid-induced diseases in agriculture. Research in virology continues to provide insights into the molecular mechanisms underlying viroid pathogenicity and the development of strategies for viroid control.

Chapter 9

Fungi and Plant Diseases

Introduction

Fungi and plant diseases play pivotal roles in shaping the health and vitality of ecosystems worldwide. This dynamic interaction between fungi and plants encompasses a complex web of relationships, ranging from mutualistic symbioses to destructive pathogenic infections. As essential components of Earth's biodiversity, fungi contribute to nutrient cycling, decomposition, and mycorrhizal associations, while plant diseases pose challenges to agriculture, forestry, and natural ecosystems.

Fungi, as a diverse kingdom of organisms, encompass a vast array of species with varied ecological functions. Many fungi form symbiotic partnerships with plants, such as mycorrhizal associations where fungal hyphae extend into the plant root system, facilitating nutrient exchange. Additionally, fungi contribute to the decomposition of organic matter, playing a crucial role in nutrient recycling within ecosystems. However, some fungi have evolved as formidable plant pathogens, causing diseases that can have profound economic and ecological impacts.

Plant diseases, caused by fungi and other pathogens, have been a constant challenge to human efforts to cultivate crops and maintain healthy vegetation. These diseases can result in reduced crop yields, loss of biodiversity, and in some cases, threaten the food security of entire populations. Understanding the mechanisms of fungal pathogenesis, host resistance, and the ecological dynamics of these interactions is essential for developing sustainable strategies to manage and mitigate the impact of plant diseases.

This exploration into the intricate relationship between fungi and plant diseases will delve into the mechanisms of fungal pathogenicity, the ecological roles of fungi in plant health, and the innovative approaches employed to control and prevent plant diseases. By unraveling the complexities of these interactions, we aim to enhance our ability to address the challenges posed by fungal pathogens and promote the health and resilience of plant ecosystems.

This is the seventh fact sheet in a series of ten designed to provide an overview of key concepts in plant pathology. Plant pathology is the study of plant disease including the reasons why plants get sick and how to control or manage healthy plants.

Fungi are spore-forming, non-chlorophytic, eukaryotic (cells having true nuclei) organisms and most of the true fungi are filamentous and branched. Most of the over 100,000 species of fungi are saprophytes. However, over 20,000 species of fungi are parasites and cause disease in crops and plants (USEPA 2005). Fungal parasites are by far the most prevalent plant pathogenic organism. All plants are attacked by one species or another of phytopathogenic fungi. Individual species of fungi can parasitize one or many different kinds of plants.

The body, or thallus, of most of the higher fungi is called a mycelium (pl. mycelia). The mycelium is made up of thread-like structures called hyphae (sing. hypha). Hyphae grow only from their tips, and under favorable conditions can grow indefinitely. Hyphae, most often, are partitioned by cross walls called septa (sing. septum). Septate hyphae are divided into individual cells by these internal walls. Septa usually have a central pore which allows small organelles, and in some cases nuclei, to pass from one cell to another. The hyphae of the classes Ascomycetes, Basidiomycetes and Deuteromycetes are septate and the hyphae of class Oomycetes and Zygomycetes are nonseptate or coenocytic. Coencytic hyphae have multiple nuclei in a common cytoplasm.

Rhizomorphs of Armillaria sp.

Hyphae produce specialized simple or branched projections called haustoria (sing. Haustorium). Haustoria penetrate the host tissue and act as nutrient absorbing organs. Haustoria contain nuclei and a concentrated number of mitochondria. The haustoria do not rupture the host cell protoplasmic membrane but invaginates itself into the cell thus greatly increasing the absorptive surface of the fungus.

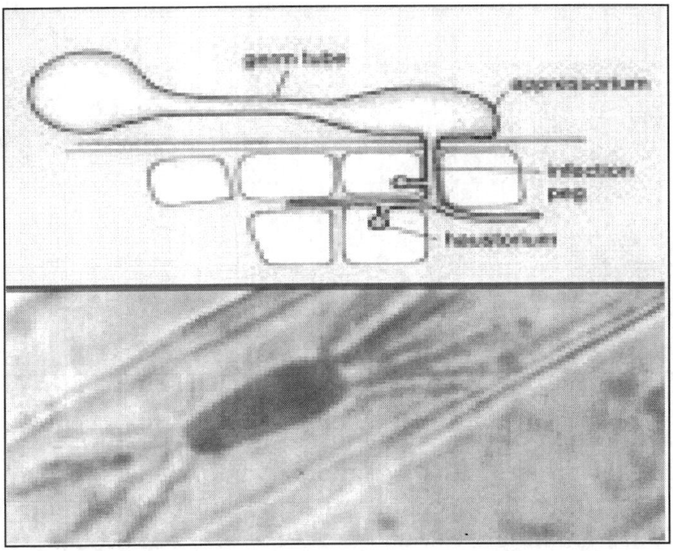

Reproduction of fungi is primarily by means of spores. Spores are reproductive bodies that consist of one or a few cells. In function they are analogous to the seeds of green plants. Spores are produced by sexual and asexual reproduction. Sexual reproduction involves the union of two compatible nuclei as produced by meiosis. For many phytopathogenic fungi the sexual cycle occurs only once during each growing season. Sexually produced spores include oospores, zygospores, ascospores and basidiospores characteristic of the phyla Oomycota, Zygomycota, Ascomycota and Basidiomycota respectively. Asexual reproduction usually occurs by means of mitosis which produce mitotic spores, mycelial fragmentation, fission and budding. With asexual reproduction the repeating cycles of infection can continue throughout the growing season. Asexual spores may be classified as oidia (formed by fragmentation of hyphae into individual cells), conidia (borne on tips or sides of specialized branches of hyphae) and sporangiospores (a nonmotile spore born in a sporangim or case).

There are many types and different characteristics of fruiting bodies, spores and mycelium. Fungi are both classified and identified by these features.

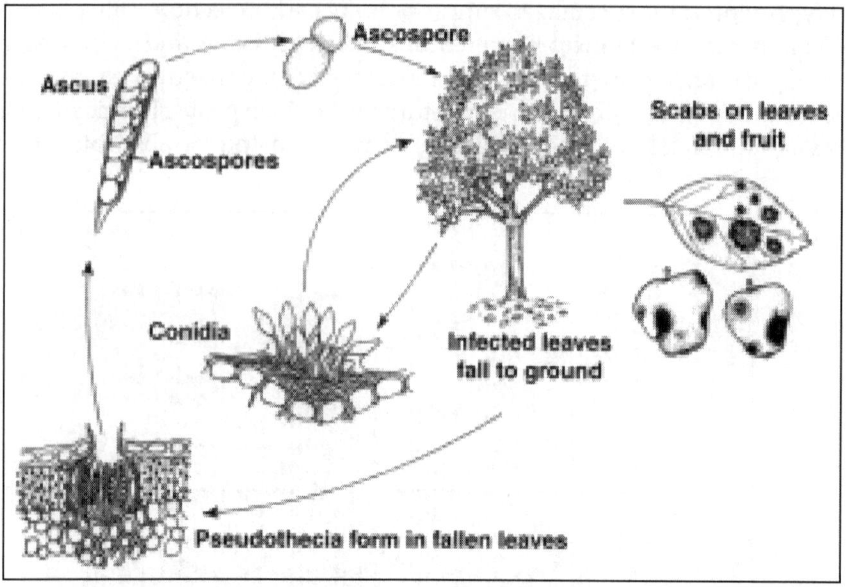

The classification of pathogenic fungi is important for identifying and diagnosing plant disease. Each kingdom, phylum, class, order, genus and specie has its own identifying characteristics. Many of the diagnostic factors in identifying different fungi are very subtle and without proper lab equipment and skill are difficult to identify down to the species level. If we can identify the type of fruiting body and/or spores that are associated with a particular phylum, we can narrow down our search considerably. Most often, we rely on visual signs and symptoms for diagnosis of fungal disease.

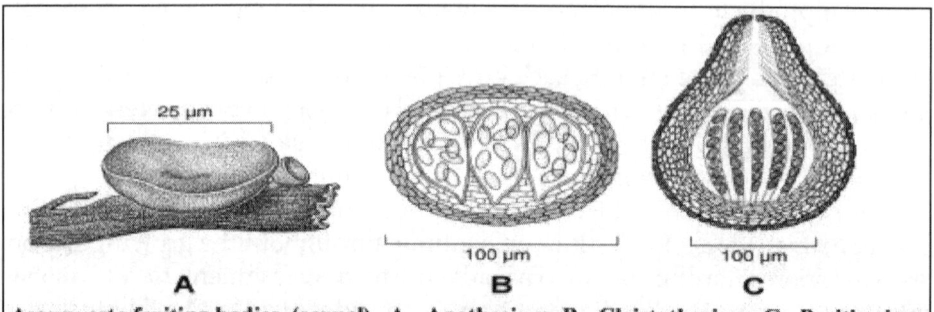

Ascomycete fruiting bodies (sexual). A - Apothecium, B - Cleistothecium, C - Perithecium

The primary phyla dealt with in plant pathology and disease diagnosis are listed below. This outline is very brief and only includes a few of the more important pathogens and problems. Also included are some of the more important characteristics of each phylum. For further clarification on fungal taxonomy, you can refer to any number of resources at the Extension office. Also, the internet is an excellent source of information on this subject.

Fungal-like Organisms (Lower Fungi)

Kingdom: Protozoa

Phylum: Plasmodiophoromycota (endoparasitic slim molds) *e.g.*: Plasmodiophora brassicae, causing club root of crucifers - Spongosphora subterranean, causing powdery scab of potato tubers.

Kingdom: Chromista

Phylum: Oomycota (biflagellate zoospores. Simple pored nonseptate hyphae. Asexual spores - sporangiospores, motile and nonmotile. Sexual spores – zoospores) *e.g.* Chrysanthemum White Rust, Phytophthora ssp. diseases, Downey Mildew.

True Fungi

Kingdom: Fungi

Phylum: Chytridiomycota (zoospores, lack true mycelium) *e.g.* parasitic on roots of cabbage and many other plants.

Phylum: Zygomycota (saprophytes or parasites of plants, (Humans and animals.) No zoospores. Simple pored nonseptate Hyphae. Asexual spores - Sexual nonmotile sporangiospores. *e.g.* Rhizopus - bread molds, rots of fruits and vegetables.

Phylum: Ascomycota (the sac fungi - most have both teleomorphic (sexual) and anamorphic (asexual) stages. Simple septate hyphae. Produce sexual spores (ascospores) or asexual spores (conidia) on free hyphae or in asexual fruiting bodies such as Cleistothecium, Perithecium, Apothecium and Pycnidium). This phylum is responsible for most of our phytopathogenic fungal diseases.

Some important pathogens in this phylum

- Peach leaf curl (*Taphrina* sp.)
- Powdery mildew (Podosphaera, (apple)
- Microsphaera (azalea) *etc.*
- Vascular wilt diseases (Verticillium, Fusarium) *etc.*

Some important signs and symptoms

- Many leaf disorders
- Fruit and vegetable rots
- Root, stem and soft rots
- Brown rot (Monilinia)
- Cankers Anthracnoses

Class: Deuteromycetes (Mitosporic or imperfect fungi-also known as Fungi Imperfecti. Sexual reproduction and structures are rare, lacking or unknown. Simple septate hyphae with some exceptions. Asexual spores - condia produced in pycnidia and acervuli).

Some examples

Asexual (anamorphic stage)	Sexual (teleomorphic stage)
Penicillium	Talaromyces
Verticillium	Hypocrea
Septoria	Mycosphaerella
Botrytis	Botryotinia
Rhizoctonia	Thanatephorus

This class is responsible for many rots, wilts, molds, leaf spots and other symptoms.

Phylum: Basidiomycota (club and mushroom fungi) - complex septate hyphae, clamp connections. Asexual spores - conidia, oidia. Sexual spores - basidiospores produced externally on club-like spore producing structures called basidium) For example:

Order: Ustilaginales - smut fungi

Corn smut

Loose smut of barley, wheat *etc.*

Bunt of wheat.

Order: Uredinales - rust fungi

Hollyhock Cedar-apple

White pine blister

Needle rusts

Cereal rusts

Order: Agaricales – mushrooms

e.g.: Armillaria spp. - root rot of fruit and forest trees.

Common signs in Armillaria are mushrooms at the soil level and mycelia under The bark of the basal portion of the tree Agaricus bisporus – common edible mushroom.

Morphology

Fungi and FLOs are eukaryotic organisms that lack chlorophyll and thus do not have the ability to photosynthesize their own food. They obtain nutrients by absorption through tiny thread-like filaments called ***hyphae*** that branch in all directions throughout a substrate. A collection of hyphae is referred to as mycelium (pl., mycelia). Mycelia are the key diagnostic sign associated with diseases caused by fungi and FLOs. Most of us have seen mycelium growing on old bread or rotten fruit or vegetables and may have referred to these organisms collectively as molds or mildew.

Fungi and FLOs (indeed all pathogens) can be grouped into the following four categories based on their preference for surviving on dead or decaying organic matter versus living tissue:

- ✰ ***Obligate saprophytes***–always a saprophyte. These organisms can only survive or are obliged to gain nourishment by colonizing dead or decaying organic matter. They are not parasites.
- ✰ ***Obligate parasites***–always a parasite. Can only grow as a parasite on or in a living host. They cannot survive as saprophytes or be cultured in the laboratory. This is a very interesting group of pathogens in that they have a vested interest in prolonging the life of their host to increase their own viability. All viruses, downy mildews, powdery mildews, rusts and smuts are obligate parasites.
- ✰ ***Facultative parasites***–usually survive as a saprophyte but have the ability to parasitize and cause disease under certain conditions. Examples include *Pythium* species and many bacterial pathogens.
- ✰ ***Facultative saprophytes***–usually survive as a parasite but have the ability to live on dead and decaying organic matter under the right conditions. Examples include *Phytophthora* and *Botrytis* species.

Some fungi and FLOs are able to live on only one host species, while others develop on many different kinds.

Diagnostic Signs and Symptoms of Fungal Infections

Detecting fungal infections in plants involves recognizing specific signs and symptoms indicative of the presence of these pathogens. Visible signs often include discoloration of leaves, wilting, and unusual growth patterns. Fungi can cause characteristic lesions, spots, or mold growth on various plant parts. Symptoms may manifest as stunted growth, distorted leaves, or premature leaf drop. Additionally, plants may exhibit a decline in overall vitality and yield. Root rot, a common fungal infection in the root system, may result in a weakened root structure and water uptake issues. Fungi can also produce distinctive spore structures, aiding in identification. Regular monitoring for these signs, coupled with timely intervention through fungicides or other preventive measures, is essential to mitigate the impact of fungal infections on plant health and agricultural productivity.

- ✰ Fungi can cause general or localized signs and/or symptoms. In the majority of cases, fungal infections cause general necrosis of host tissue and often cause stunting, distortions and abnormal changes in plant tissue and organs.

☆ The most distinctive and easily identifiable characteristics of fungal infections are the physical presence of signs of the pathogen. Signs include hyphae, mycelia, fruiting bodies and spores of the fungal pathogen are significant clues to proper identification and diagnosis of a disease. The fruiting bodies of fungi range from microscopic to macroscopic. They come in many shapes and configurations and have their individual characteristics. The fruiting bodies, along with spores, and mycelium, in most cases can lead to an accurate identification of the disease.

☆ The following symptoms are common in fungal infections whether alone or in combination with other fungal pathogens.

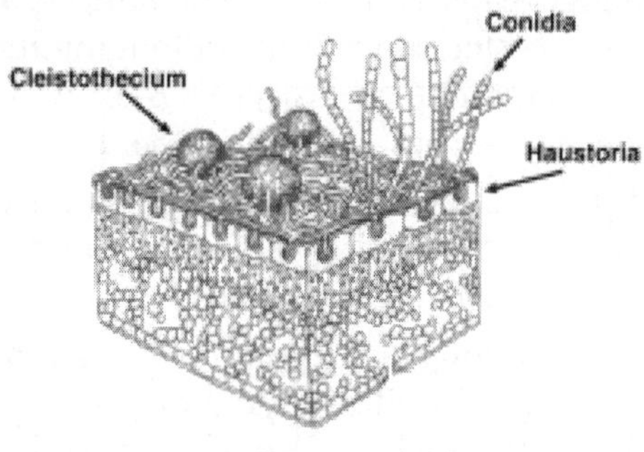

Powdery mildew on leaf surface

☆ Leaf Spots are very common in both biotic and abiotic plant disorders. Fungal leaf spots often take the form of localized lesions consisting of necrotic and collapsed tissue. Leaf spots can vary in size and are generally round and concentric, but can be ovoid or elongated on both leaves and stems of the host. The typical fungal leaf spot will have a "bulls-eye-like" appearance consisting of roughly concentric rings that may display zones of different colors such as yellow, red or purple, and will often have a tan center. As the spots develop, they are not restricted by the leaf veins as can be the case in bacterial leaf spots. Fungal leaf spots will usually have a dry texture but are not dry and papery.

Common Symptoms of Fungal Infections in Plants

Plant fungal infections can have detrimental effects on crop health and agricultural productivity, making early detection crucial for effective management. Recognizing the common symptoms associated with these infections is key to implementing timely interventions. Fungi can affect various parts of the plant, manifesting in visual cues such as discoloration, wilting, and abnormal growth patterns. Leaves may exhibit characteristic lesions, spots, or mold growth, while the overall vigor and yield of the plant may decline. Root rot, another prevalent fungal infection, can compromise the root system, leading to weakened structures and impaired water uptake. This introduction sets the stage for understanding the signs that signal the presence of fungal infections in plants, facilitating proactive measures to safeguard crop health and ensure sustainable agriculture.

Anthracnose

An ulcer-like lesion that can be necrotic and sunken. These lesions can appear on the fruit, flowers and stems of the host - *e.g.* Apple Anthracnose of stems and or leaves (*Cryptosporiopsis* sp. Formally *Pezicula* sp.), or Dogwood Anthracnose (*Discula distructiva*).

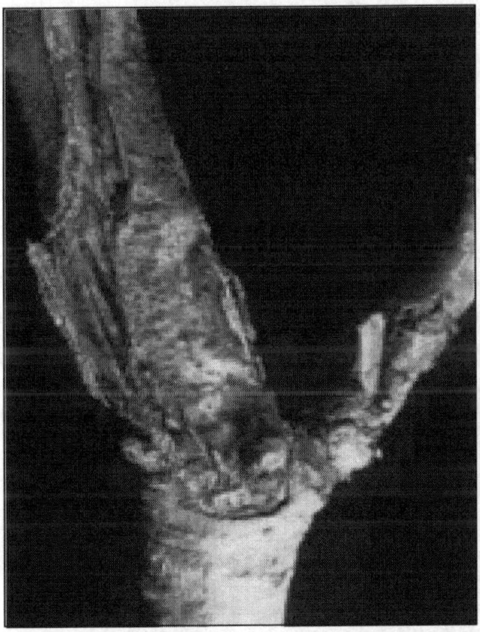

Apple Antrhacnose

Canker

Canker

A localized necrotic lesion on woody tissue, often sunken - *e.g.* Apple European Canker (*Nectria galligena*), Douglas fir, upper Twig Canker (*Phomopsis* sp.), and many more.

Damping Off

A rapid collapse and death of very young seedling. Either the seed rots before emergence or the seedling rots at the soil line and falls over and dies. Several soil-born fungi cause this disease. The most common genera involved are *Fusarium, Rhizoctonia* and *Pythium*.

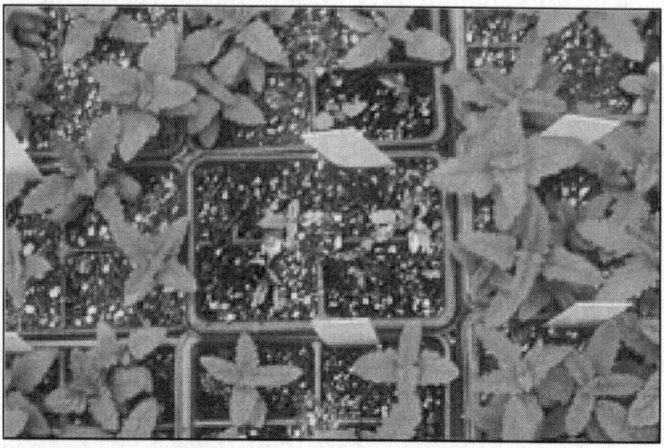

Damping Off of seedlings

Scab

Localized lesion on host fruit leaves tubers and other plant parts. These infections usually result in a roughened, crust-like area on the surface of the host - *e.g.* Apple Scab (*Venturia inaequalis*) and Pear Scab (*Venturia pirina*).

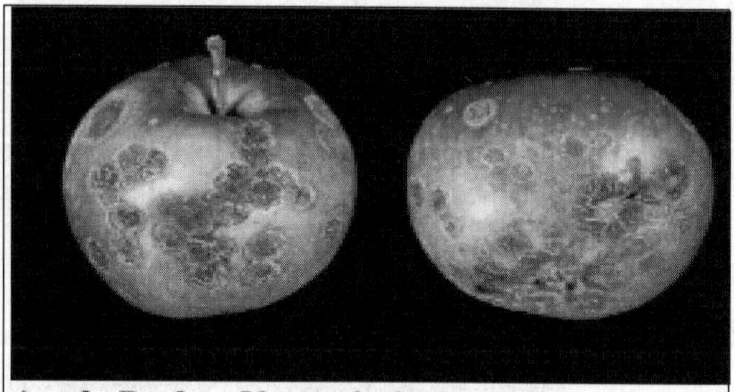

Apple Scab – *Venturia inaequalis*

Soft and Dry Root Rots

Rot and disintegration of fleshy leaves, roots, tubers and fruit - *e.g.* Damping Off of seedlings, Rhododendron Phytophthora Root Rot (*P. cinnamomi*), *Armaillaria* spp. Root Rot of trees and hundreds of other plants.

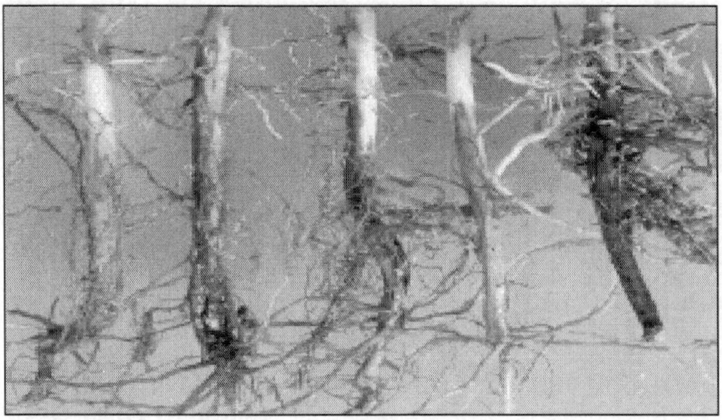

Blight

Rapid generalized browning and death of leaves, floral organs, stems and branches. Blights can refer to both biotic and abiotic disorders - *e.g.* Cherry Brown Rot Blossom Blight (*Monilinia fructcola* and *M. laxa*), Tomato Early Blight (*Alternaria tomatophilai*, formally, A. solani) and Tomato and Potato Late Blight (*Phytophthora infastans*). *P. infestans* was responsible for the great potato famine in Ireland in 1845.

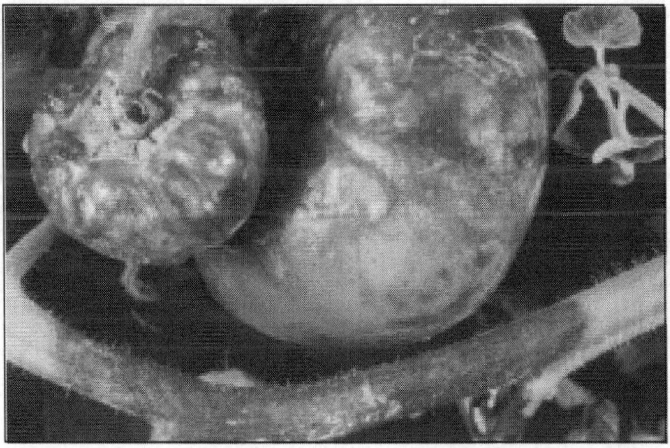

Dieback

Progressive death of shoots and twigs generally starting at the tip of the infected plant part - *e.g.* Shoot Dieback of Apple due to Brown Rot of Cherry (*Monilinia* sp.) and Poplar Shoot Dieback (*Venturia populina*). Bacteria are probably more commonly associated with Diebacks (*Pseudomonas syringae*).

Decline

Progressive loss of vigor over a period of time. General decline can be caused by many factors including fungal, bacterial, environmental and other factors, usually in combination.

Phytophthora Root Rot

Other symptoms associated with excessive growth (hyperplasia) or enlargement (hypertrophy) and distortion of plant parts can be demonstrated by the following:

Galls

Enlarged parts of plant organs, usually caused by excessive multiplication or enlargement of plant cells - *e.g.* Camellia Leaf Gall (*Exobasidium camelliae*), Plum and Prune Black knot (*Apiosporina morbosa*), Pine Western Gall Rust (*Peridermium harknesii*), Clubroot - (*Plasmodiophora brassicae*) enlarged roots that look like clubs or spindles - *e.g.* Clubroot of Crucifers (*Plasmodiophora brassicae*), Burr Knot of apples caused by environmental and/or genetic factors can be similar to bacterial galls.

Leaf Curls

Curling, thickening and distortion of leaves - *e.g.* Almond Leaf Curl, Peach Leaf Curl, Pear Leaf Blister, Maple Leaf Curl and many more caused by Taphrina sp.

Other Fungal Symptoms

Wilts

Generalized loss of turgidity as in vascular wilts – Maple Verticillium Wilt, Damping Off and so on. Both woody and herbaceous plants are subject to wilts.

Maple Verticillium Wilt - Verticillium sp.

Mildews

Mycelium, fruiting bodies and necrotic tissue. *e.g.* Powdery Mildews, Downey Mildews.

Powdery Mildew

Rusts

Infected plants will most of the time have many small lesions on stems or leaves, usually a rust color but can also be black or white.

Smuts

Mycelium or black spores on seeds, in the form of galls or seeds "replaced" by spores.

Beneficial Fungi

Fungi and FLOs can be beneficial as well as pathogenic. Beneficial fungi participate in biological cycles such as decaying dead animal and plant materials converting them into nutrients that are absorbed by living plants. Some beneficial fungi grow in a symbiotic relationship with the root cells of higher green plants; this life style is termed ***mycorrhizal***. Roots of most cultivated plants–corn, soybeans, cotton, tobacco, peas, red clover, apples, citrus, pines, aspens, birches, turfgrass species and others–have mycorrhizal relationships with soil fungi. The mycorrhizae appear to be highly beneficial, often necessary, for optimum growth of many plants. Some beneficial fungi, such as those belonging to the genus *Trichoderma*, are effective biocontrol agents of plant pathogenic fungi while others, like *Arthrobotrys dactyloides*, have been shown to trap and parasitize plant pathogenic nematodes.

Certain fungi produce useful antibiotics and enzymes. *Penicillium* fungi produce the famous penicillin G, which has prevented countless deaths from bacterial infection, acting by inhibiting formation of the bacteria's cell wall. Many food-producing processes, such as the making of bread, wine, beer and cheese, are based on the activities of fungi. Most notably, mushrooms, which are fungi, are an important food for humans, animals and insects.

Pathogen Biology

Because of the shear number of plant diseases caused by fungi and the huge diversity in how plant pathogenic fungi cause disease, it is impossible and beyond the scope of this publication to provide details about specific disease cycles and integrated fungal disease management strategies. But similar to all other groups of plant pathogens, fungal pathogens have developed ways to survive periods of unfavorable environmental conditions or in the absence of a susceptible host, spread, infect, grow and reproduce on and within plants. One important difference between fungi and FLOs vs. bacteria and viruses is fungi and FLOs can penetrate a host via a wound or natural opening, but they can also actively penetrate via the production of specialized hyphal structures called appresoria (sing., appresorium). Appresoria are swollen tips of hyphae that allow the fungus, through mechanical and enzymatic activity, to directly penetrate plant tissues.

Phytophthora Root and Stem Rot of Soybean.

Brown Rot of a Peach Fruit.
Photo courtesy D.F. Ritchie, © The American Phytopathological Society.

Control Measures

There are generally more options available to professional plant production specialists and growers to manage fungal and FLO diseases as compared to viral and bacterial diseases. One of the most satisfactory methods of dealing with fungal diseases is strict sanitation to eliminate the pathogenic organism, starting with the initial stages of propagation and growth of the potential host plants. Integrated management strategies for fungal and FLO diseases include the following:

Genetic Host Resistance

Using genetically resistant species, cultivars, varieties and hybrids. In many of the major crops, cultivars resistant to prevailing diseases are available, and more are continually being developed by plant breeders. As has been discussed with other types of diseases, the use of genetically resistant plants, if available, should be the first line of defense for diseases caused by fungi and FLOs. Notable examples include: Certain hybrid potato cultivars are resistant to late blight (*Phytophthora infestans*); Soybean cultivars resistant to downy mildew (*Peronospora manshurica*) have been developed; In the United States, apple cultivars are available from the Indiana and the New York agricultural experiment stations that show high resistance or immunity to apple scab (*Venturia inaequalis*); In the cereal crops (oats, wheat, rye, barley), powdery mildew (*Erysiphe graminis*) can be controlled only by the use of resistant cultivars developed by plant breeders; Tomatoes can be grown in *Fusarium*-infested soils only if *Fusarium*-resistant cultivars are planted; Plant breeders are continuously developing wheat cultivars resistant to stem rust (*Puccinia graminis tritici*), but the fungus rapidly mutates, attacking the formerly resistant cultivars, requiring new resistant cultivars to be developed.

Cultural Practices

Planting only disease-free certified seed. Maintaining a balanced fertility program that avoids excessive or inadequate levels of key plant nutrients. Maintaining an effective water management program–maintain adequate soil drainage, monitor irrigation practices, and adjust accordingly, *etc.* Ensuring proper lighting–both quality and quantity to optimize plant health–especially important in turfgrass, floricultural and ornamental nursery production systems. Removing crop residues by burning or burying (plowing). Implementing crop rotation strategies to reduce or eliminate the interaction of susceptible plants with pathogens. Growing crops in climates unsuitable for pathogenic fungi and FLOs. Careful handling of the crop (vegetables and fruits) to prevent cuts, bruises and wounding during harvest, transit and storage. Storage of crop products at the proper temperatures. Soil pasteurization (moist heat at 82°C for 30 minutes).

Chemical Applications

The use of preplant soil fumigants, the use of fungicide drenches or seed treatments with fungicides, Fungicide applications, Postharvest treatment of fruits and vegetables with fungicides. Although the use of resistant cultivars and eradication of the pathogen through the use of cultural practices are the most satisfactory ways of dealing with diseases caused by fungi and FLOs, in many instances these measures are not possible. Often the disease appears and its development must be slowed or stopped by whatever means are available. Fungicide applications are often essential where there is a demand for plant health during environmental periods that favor pathogen growth. They are typically more effective when applied prior to the onset of disease symptoms (referred to as preventive or preventative applications). Some fungicides are effective when applied after the onset of symptoms and are said to have curative activity. In either case, fungicides must be delivered to the area of the plant where the pathogen is active to be effective. There are many different types and chemical classes of fungicides currently available. Numerous online extension-outreach and agrichemical company resources exist that provide specific fungicide recommendations for nearly every major cropping system and pathogen. Always read and follow label recommendations when applying pesticides.

Biological Control

Biological control serves as an environmentally friendly and sustainable approach to managing fungal plant diseases. In this method, living organisms, such as beneficial microbes or predators, are utilized to suppress or control the growth of pathogenic fungi. One common strategy involves the introduction of antagonistic microorganisms, like certain strains of bacteria or fungi, which compete with and inhibit the growth of plant pathogens. Additionally, the use of beneficial insects or nematodes that prey on fungal pests contributes to the natural balance within ecosystems. This approach not only minimizes the reliance on chemical fungicides but also fosters long-term disease management by promoting ecological equilibrium. By harnessing the power of nature's own checks and balances, biological control emerges as a sustainable and ecologically sound solution to combat fungal plant diseases in agriculture. The use of biological control organisms to suppress the activity of deleterious fungi and FLOs.

Government Regulatory Measures

Government regulatory measures play a vital role in controlling fungal plant diseases to ensure the health of crops and prevent economic losses in agriculture. Regulatory frameworks are designed to monitor, prevent, and manage the spread of fungal pathogens. These measures typically include:

- **Quarantine and Inspection Protocols:** Governments establish quarantine regulations to restrict the movement of plants, plant products, and soil that could potentially harbor fungal diseases. Inspection protocols at borders and within countries help identify and prevent the introduction or spread of infected plant materials.
- **Certification Programs:** Certification programs are implemented to ensure that plants and plant products meet specific health standards. Certified disease-free seeds and plants help prevent the introduction of fungal pathogens into new regions.
- **Fungicide Registration and Use Guidelines:** Governments regulate the registration and use of fungicides to ensure their effectiveness, safety, and minimal environmental impact. Guidelines help farmers make informed decisions about the application of fungicides.
- **Research and Surveillance Programs:** Governments invest in research and surveillance to monitor the prevalence of fungal plant diseases. Early detection and tracking of emerging threats allow for prompt responses and the development of effective control strategies.
- **Public Awareness and Education:** Government agencies often engage in public awareness campaigns and educational programs to inform farmers, agricultural professionals, and the general public about the risks associated with fungal plant diseases. Training initiatives promote proper disease identification and management practices.
- **International Cooperation:** Governments collaborate with international organizations to exchange information, adopt best practices, and coordinate efforts to address cross-border challenges

related to fungal plant diseases. This global cooperation helps prevent the spread of pathogens across regions.

- By implementing these regulatory measures, governments contribute to the overall biosecurity of their agricultural systems, safeguarding crops and ensuring a stable and secure food supply.

Chapter 10

Nematodes and Plant Diseases

Introduction: An Overview

Nematodes, commonly known as roundworms, constitute a diverse and abundant group of microscopic organisms that play crucial roles in various ecosystems. While many nematode species are beneficial and contribute to nutrient cycling and soil health, a subset of these microscopic worms can pose significant threats to plants, leading to various plant diseases. The interactions between nematodes and plants form a complex web that influences agricultural productivity and ecosystem dynamics.

Plant-parasitic nematodes are microscopic worms that feed on plant tissues, primarily roots, causing a range of diseases known as nematode-induced plant diseases. These nematodes belong to different families, such as Meloidogynidae (root-knot nematodes), Heteroderidae (cyst nematodes), and Tylenchidae (lesion nematodes), among others. Each family has distinct characteristics and life cycles, contributing to their varied effects on plant health.

The impact of nematodes on plants is multifaceted. They can directly damage plant tissues by feeding on root cells, leading to stunted growth, reduced nutrient uptake, and increased susceptibility to other pathogens. Additionally, nematodes can serve as vectors for plant viruses, further complicating disease dynamics in agricultural and natural ecosystems.

The management of nematode-induced plant diseases poses a significant challenge for farmers and researchers alike. Traditional approaches involve the use of chemical nematicides, but environmental concerns and the development of nematode resistance have prompted the exploration of alternative strategies. Integrated pest management (IPM) practices, including crop rotation, resistant cultivars, and biological control agents, offer sustainable solutions to mitigate nematode impact while preserving soil health.

Understanding the intricate relationships between nematodes and plants is essential for devising effective strategies to manage plant diseases caused by these microscopic organisms. Ongoing research in nematology aims to unravel the molecular and ecological aspects of nematode-plant interactions, providing valuable insights for the development of innovative and sustainable approaches to safeguard global agriculture. As we delve deeper into the complexities of nematode biology and their impact on plant health, we move closer to a future where resilient crops coexist harmoniously with these tiny yet influential organisms.

A number of genera and species of nematodes are highly damaging to a great range of hosts, including foliage plants, agronomic and vegetable crops, fruit and nut trees, turfgrass, and forest trees. Some of the most damaging nematodes are: Root knot (*Meloidogyne* spp.); Cyst (*Heterodera* and *Globodera* spp.); Root lesion (*Pratylenchus* spp.); Spiral (*Helicotylenchus* spp.); Burrowing (*Radopholus similis*); Bulb and stem (*Ditylenchus dipsaci*); Reniform (*Rotylenchulus reniformis*); Dagger (*Xiphinema* spp.); Bud and leaf (*Aphelenchoides* spp.); and Pine Wilt Disease (*Bursaphelenchus xylophilus*).

Morphology of Nematodes

Nematodes, also known as roundworms, exhibit a simple yet diverse morphology. Their structure is well-adapted for a wide range of environments, and they can be found in various habitats, including soil, water, and the tissues of plants and animals. Here are the key aspects of nematode morphology:

- **Body Shape:** Nematodes have a long, cylindrical body that is tapered at both ends, resembling a slender tube. The body is unsegmented and lacks external appendages.
- **Cuticle:** The outermost layer of the nematode's body is called the cuticle. It is a tough, flexible, and protective exoskeleton that covers the entire body. The cuticle is molted periodically as the nematode grows.
- **Body Wall Layers:** Beneath the cuticle, the body wall consists of three layers: the outer hypodermis, the middle muscle layer, and the inner pseudocoelom. The pseudocoelom is a fluid-filled body cavity that provides structural support.
- **Head Region:** The anterior end of the nematode is the head region. It often contains sensory organs, including amphids (ciliary sense organs), papillae, and sensory bristles.

- **Mouth and Buccal Cavity:** Nematodes have a simple, tubular mouth leading to a muscular buccal cavity. The buccal cavity may contain structures such as stylets or teeth, which are used for feeding on plant or animal tissues.
- **Digestive System:** The digestive system of nematodes is relatively simple, consisting of a mouth, a muscular pharynx, an intestine, and an anus. The pharynx is responsible for ingesting food particles, and some nematodes have specialized structures for piercing and sucking plant or animal tissues.
- **Reproductive System:** Nematodes are dioecious, meaning there are separate male and female individuals. The reproductive system includes gonads, which produce eggs in females and sperm in males. Mating involves the transfer of sperm from males to females.
- **Excretory System:** Nematodes possess a simple excretory system known as the excretory pore or "renette cell." It helps regulate the osmotic balance within the nematode's body.
- **Lateral Cord and Nerve Ring:** Nematodes have a pair of lateral nerve cords running along the length of the body. These cords are connected by a nerve ring encircling the pharynx, forming part of the nervous system.
- **Tail Region:** The posterior end of the nematode is the tail region. The tail varies in shape and may end in a pointed tip or a more rounded terminus.

Understanding the morphology of nematodes is crucial for the identification of different species and for studying their biology, ecology, and potential impact on plants, animals, and humans.

- Nematodes are simple, multi-cellular animals–typically containing 1,000 cells or less. They are worm-like in appearance, but are taxonomically distinct from earthworms, wireworms or flatworms. They are bilaterally symmetrical, soft-bodied (no skeleton), non-segmented round worms. Most nematode species that attack plants are microscopic. The basic body plan of a nematode is a "tube within a tube." Nematodes feed on other microorganisms and plants like bacteriovores, fungivores, omnivores, predators, and plant parasites. Some, however, are serious human, animal and plant pathogens. Those that attack animals or humans do not attack plants and vice versa. Heartworm in dogs and cats is an example of nematode diseases in animals and people.
- Plant parasitic nematodes may attack the roots, stem, foliage and flowers of plants. All plant parasitic nematodes have piercing mouthparts called stylets. The presence of a stylet is the key diagnostic

sign differentiating plant parasitic nematodes from all other types of nematodes. The bacterial-feeding nematode, Caenorhabditis elegans, is one of the best-understood animals on earth. It was the first animal to have its entire genome completely sequenced. The study of C. elegans has led to many new insights into animal development, neurobiology and behavior.

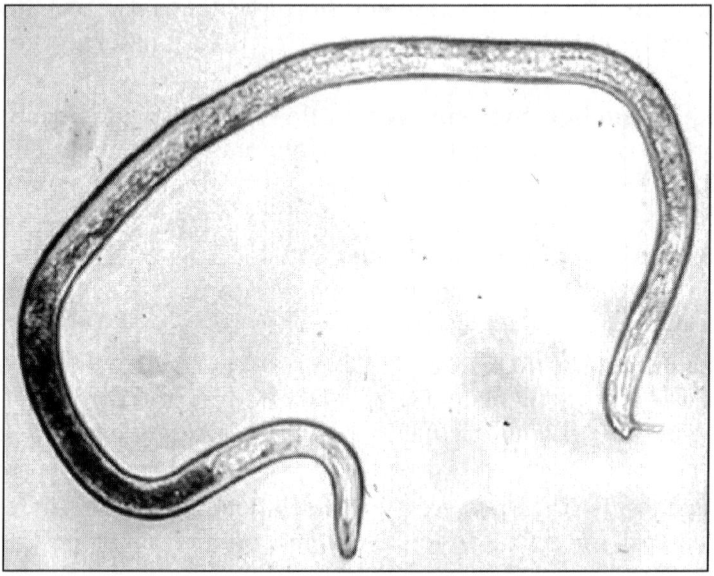

Figure 10.1: Adult Root-knot Nematode.

Common Signs and Symptoms of Plant Nematodes

Plant nematodes can cause a variety of symptoms, and the signs of their presence may not always be immediately visible. Nematode damage can be subtle and gradual, often resembling symptoms caused by other factors such as nutrient deficiencies or diseases. Here are common signs and symptoms associated with plant nematodes:

- ☆ **Stunted Growth:** Infected plants may exhibit reduced overall growth, including shorter stature and smaller leaves. Nematodes can affect the plant's ability to take up nutrients, leading to stunted development.
- ☆ **Yellowing (Chlorosis):** Nematode-infested plants often show symptoms of yellowing, especially in older leaves. This can be attributed to the nematodes disrupting the plant's vascular system and impeding nutrient transport.
- ☆ **Wilting:** Nematodes may induce wilting in plants by causing damage to the root system. The compromised roots are less efficient in water uptake, leading to wilting even when soil moisture is adequate.

- **Root Galls or Swellings:** Root-knot nematodes (*Meloidogyne* spp.) induce the formation of galls or swellings on the roots. These abnormal growths can interfere with nutrient and water absorption, contributing to overall plant decline.
- **Root Lesions:** Lesion nematodes (*Pratylenchus* spp.) feed on plant roots, causing necrotic lesions. Infected roots may appear discolored, and the lesions can serve as entry points for secondary infections.
- **Root Rot:** Nematodes, particularly lesion nematodes, can make plants more susceptible to root rot diseases by creating wounds on the roots. This can lead to further decay and reduced root function.
- **Reduced Yield:** Nematode-infected plants often experience a decrease in yield, as the compromised root system hinders the plant's ability to take up nutrients and water essential for reproductive processes.
- **Patchy or Uneven Crop Growth:** Nematode damage may result in irregular patterns of growth within a field or garden. Some areas may have healthier plants, while others exhibit signs of stress and decline.
- **Secondary Infections:** Weakened plants are more susceptible to secondary infections by bacteria, fungi, or other pathogens. Symptoms of these secondary infections may mask the primary nematode-induced damage.
- **Visible Nematode Infestations:** In some cases, nematodes themselves may be visible on the roots or in the soil around the root zone, especially during severe infestations. However, microscopic examination is often required for accurate identification.

It's important to note that nematode symptoms can vary depending on the type of nematode, the host plant, and environmental conditions. Early detection and accurate diagnosis are crucial for implementing effective management strategies to mitigate the impact of nematodes on plant health.

Dissemination

- Parasitic nematodes are readily spread by any physical means that can move soil particles about–equipment, tools, shoes, birds, insects, dust, wind and water.
- In addition, the movement of nematode-infested plants or plant parts will spread the parasites.

Control Methods

The following points highlight the six main types of management of nematodes pests. The types are:

- Cultural Management

- ☆ Biological Management
- ☆ Regulatory
- ☆ Regulatory Management
- ☆ Integrated Management of Nematodes

Cultural Management

The affectivity of cultural method depends upon the proper knowledge of life history, population dynamics host range of the plant parasitic nematodes infesting crops. The cultural practices include crop rotation, fallowing, flooding, sanitation, ploughing during summer season, mulching, organic manure, spacing of plants in the field, time of sowing, resistant varieties *etc.* Nematode management through inter cropping and crop rotation is based on the fact that some species of nematodes are able to feed and multiply only on host crop. Rotation is a very old practice for reducing soil borne problems. Rotation to non-host crops may cause many of those pests to cease reproduction and allow natural mortality factors to reduce their numbers. Flooding or fallowing may be used to help reduce numbers of nematode pests. The period of flooding appears to vary with several factors such as kind of soil, season *etc.* Warm conditions are said to reduce the period of time required for control. Juveniles (larvae) are more easily killed by flooding than eggs. The period of flooding needs to be worked out for each condition. Fallow periods in cropping sequences can also reduce nematode populations. The summer ploughing 2-3 times during hottest period of the year help to expose nematodes to the drying action of sun and wind and reduce the population. Haque and Prasad (1982) reported drastic reduction of plant parasitic nematodes and significantly increased the yield with three deep ploughing. The intensity of symptoms was directly related to the inoculum levels. Sanitation terms covers a wide range of cultural practices, including weed control, crop residue destruction and disinfestations of farm equipment before moving it from heavily infested fields to uninfected fields. In monocultures, eliminating the weed hosts can be important in reducing the populations of plant parasitic nematodes. Soil temperature plays crucial role in the activities of plant parasitic nematodes, the time during which crop is planted is important. The addition of inorganic fertilizers alone without organic manure usually increases the nematode population and disease intensity. NH4-N reduces the disease incidence while NO3-N may increase the same. Particular forms as well as dose and proportion of NPK may also reduce or increase the incidence of disease. Use of tolerant/ resistant varieties is most practical approach for the management of nematode diseases. Crop cultivars resistant to phytonematodes can be the most useful and cheapest means of nematode control for the small-scale farmers. Nematicidal plants with roots containing nematicidal substances have been investigated. These toxic substances reduce the population level of some nematode species. African marigolds (Tagetes spp.), asparagus, crotalaria, mustard and several cruciferous plants have been reported to produce toxic substances. Soil amendments with green manure, compost, oil cakes of neem, mahua, mustard, groundnut, cotton, linseed, karanj and saw dust *etc.* have been found to reduce nematode populations. Neem, karanj and groundnut cakes incorporated into soil at the rate of 1,500 kg/ha give good control of plant parasitic nematodes and could be practiced wherever possible. Apart from encouraging the multiplication of natural enemies like nematode trapping fungi, the decomposition products of these organic amendments are toxic to nematodes. Use of organic manures is of great value since the decomposition products and promotion of natural enemies decrease nematode populations. The amount of oil cakes or any organic matter to be incorporated depends on various factors like, soil type and texture, crop to be planted, the predominant nematode fauna of the soil and the amount of soil moisture present in the soil. In trap crops when grown in infested soils, the nematodes penetrate into the root system and start multiplying. Before the nematodes complete its life cycle the plants are uprooted and destroyed. Applications of botanicals are easy, environmentally safe, and having no phytotoxic effect on crops. The inhibition of root-knot development may be due to the accumulation of toxic by products of decomposition to increased phenolic contents resulting in host resistance, or to changed physical and chemical properties of soil inimical to the nematodes. One of the most economical and effective ways to control plant parasitic nematodes is the growing of nematode resistant plant cultivars. The identification of a dominant Mi gene importing resistance to the root- knot nematodes and its linkage to an acid phosphates gene is a major break through in this area of research.

Biological Management

Use of biocontrol agents in nematode management of vegetables are increasing. Application of natural enemies of plant parasitic nematodes for controlling nematode population is an essential component of eco-friendly management. Nematologists in all over the world are working very hard to identify and lean to manipulate natural enemies of nematodes so they can be used as biological control agents. Nematodes have many natural enemies including fungi, bacteria and predacious nematodes. Certain fungi capture and kill nematodes in the soil. *Arthrobotrys* spp., *Dactylaria* spp., *Dactylella* spp., *Catenaria* spp., and *Trichothecium* spp., are the genera most commonly represented. The biological control of root knot nematode on tomato under green-house condition by using predaceous fungi has been reported Singh 2001; Bandyopadhya 2001. Some fungi capture nematode by adhesion, but many employ specialized devices that include networks of adhesive branches, stalked adhesive knobs, non-constricting rings and constricting rings. The surface of the nematode is penetrated and the fungus hyphae grow throughout the nematode body, digesting and absorbing its contents. Under favorable conditions, large numbers of nematodes may be captured, and killed especially by those fungi that form adhesive net-works or hyphal loops. *Trichoderma harzianum, Trichoderma virens, Aspergillus niger, Paecilomyces lilacinus, Pochonia chlamydosporia* are found promising biocontrol agents. Davide and Zorilla (1985) while investigating the biocontrol potential of P. lilacinus against M.incognita on okra found the fungus to be quite effective and economically better than nematicides. According to Shahzad and Ghaffar (1984) carbofuran at the rate 1 kg a.i/ha was less effective than *P. lilacinus* against *M. incognita*. Now mycorrhiza is not restricted to its use only as biofertilizers, its potential role in the biological control of plant parasitic nematodes is reported by many workers. Sikora (1979) found that prior presence of VAM fungi Glomus mosseae has resulted into an increase in plant resistance against *Meloidogyne* spp. A bacterial parasite of nematodes *Pasteuria penetrans*, has received much attention and research effort in recent years, *P. penetrans* is probably the most specific obligate parasite of nematodes, with a life cycle remarkably well adapted to parasitism of certain phytonematodes. It directly parasitizes juvenile nematode, thus affects penetration and reproduction. Pasteuria penetrans can survive several years in air dried soil apparently without loss of viability. Seed bacterization, soil drenching and bare root dip application with *Pseudomonas fluorescens, Pasteuria penetrans, Bacillus subtilis, B. polymyxa* effectively controls plant parasitic nematodes was also reported by many workers. Among the predatory nematodes, monarchs may be proved efficient predators because of stronger predatory potential, high rate of predacity and high strike rate.

Chemical Management

Many of the nematicides are broad spectrum volatile soil fumigants that are active against not only nematodes but also insects, fungi and bacteria living in the soil. Two major groups of nematicides are distinguished by the manner in which they spread through the soil. Soil fumigants are gases in cylinders or liquids which spread as gases from the point where they are infected into the soil. Non fumigant nematicides include a variety of water soluble compounds which are applied to the soil as liquid or granular formulations. Most belong to the carbamate or organophosphate families of pesticides. These distributions in the soil depend on physical mixing during application and moving in solution in soil water. There are no perfect nematicides for all purposes. Nematicides vary in their effectiveness against different kinds of nematodes, ease to handling, cost effect on other classes of pests (weeds, disease organisms, and insects) behavior in different soils, toxicity to different plants and availability. The performance of the nematicides will depend on soil conditions, temperatures and rainfall. A yield benefit is not guaranteed and nematicides are expensive. The older nematicides are mostly fumigants which are applied in soil. These are seldom recommended because of their hazardous nature and high toxicity to non-target organisms. Recently, large number of non-fumigant and systehiic nematicides are available which are safe on plant. DBCP (nemagon) has been found very effective for standing crops against root-knot nematodes but its use has been suspended due to adverse effect on human beings. Carbofuran

3G (furadon), phorate 10G (thimet), fenamiphos (nemacur), fensulphothion (dasanit) have been recommended for a variety of vegetable crops by several workers. The development of the *H. cajani* nematode was also delayed in pot condition experiment which was treated with carbofuran, phorate and aldicarb nematicides. The dose and method of application would vary with crop. Chemicals and Phytoalexins that have implications in the nematode control, synthetic pheromones to disrupt nematode reproduction and chemicals that directly interfere with nematode chemoreceptors should be exploited for nematode population management. Prasad (1990) studied the control by seed pelleted with carbofuran, aldicarb, sulfone at the rate 3 and 6 per cent w/w. Application of carbosulfan nematicides reduced root-knot nematodes in vegetables and seed treatment was more effective than soil application.

Regulatory Management

Numerous attempts have been made to prevent the introduction of nematodes into countries or provinces by means of quarantine. Quarantines are established by legislative action in parliament, *etc.*, and usually give quarantine authorities power to make and enforce regulations to accomplish the purpose. Such regulations usually prohibit bringing infected seeds into protected areas where similar crops might become infected.

Physical Management

The scientific principle involved in thermotherapy is that the pathogens present in soil and plant materials are inactivated or eliminated at temperatures non-lethal for the host tissues. Physical means of nematode control includes heat treatment of soil, solar drying, steam sterilization, hot water treatment and soil solarization. Soil solarization with transparent polyethylene sheet has been attended as a means of raising the soil temperature to lethal levels to control soil pathogens. Soil solarization, a method of pasteurization can effectively suppress most species of nematode along with other microbes and weeds under field conditions. Generally, sheets of 50-100 pm thickness are most suitable for raising the soil temperature. The use of transparent polyethylene had yielded better results than black sheets, since transparent sheet transmit most of the incident radiation to soil. Additions of salt, sugar, charcoal *etc.* create osmotic stress on nematodes and can be used for controlling nematodes.

Integrated Management of Nematodes

Integrated nematode management is based upon the system approach, follows location specific principle and is environment specific. Utilization of the best combination of available management strategies for the pest complex at hand (nematodes, insect pests, disease organisms, weeds *etc.*) constitutes an integrated crop protection system. Resistant-cultivars, crop rotation, pesticides, and sanitary and cultural practices can all be employed to the best possible advantage. An integrated management strategy prevents the excessive build up any single nematode, insect, or disease population and minimizes the development of pest resistance to any single tactic. Integrated pest management systems require flexibility and depend upon the specific pest problem and locally available management options. A fixed set of recommendations may keep a pest complex in check for a limited period of time, but as the pest population shifts, recommendations will have to change also. Therefore, system development takes into account many factors including the species and race of pests present, the availability of resistant host plants, the longevity of the pest, and the crops, cropping systems, and climate of the geographical region. The end result is a management strategy tailored to fit the unique circumstances of each pest situation.

Chapter 11

Phanerogamic Plant Parasites

Introduction

It seems like there might be a misunderstanding in your request. The term "phanerogamic" typically refers to seed-bearing plants, and "parasites" are organisms that live on or in another organism (the host) and derive nutrients at the expense of the host. If you're looking for information on parasitic seed-bearing plants, I can provide some general information.

One example of a parasitic seed-bearing plant is the dodder (genus Cuscuta). Dodder is a genus of about 200–170 species of parasitic flowering plants. These plants lack chlorophyll and cannot photosynthesize, so they rely entirely on their host plants for nutrients. Dodder seeds germinate near a host plant, and the dodder vine then wraps itself around the host, penetrating its vascular tissue to extract water and nutrients.

General Properties of Parasitic Plants

- ☆ Nutrients and water are transported via a Physiological bridge called the Haustorium.
- ☆ A parasite connects its vascular system (at least one of the tissues) to that of the host plant.
- ☆ The parasite may totally discard its own photosynthesis.
- ☆ Parasites may be mostly exposed at the surface of the host (Epiparasite).
- ☆ Mostly parasite hidden within the host organ (Endoparasite).
- ☆ The endoparasitic portion is composed of thread-like haustoria permeating the host tissue with a sinker, a single structure that becomes embedded in the host tissue.

Types of Phanerogamic Parasites

Phanerogamic parasites, also known as higher plants that are parasitic, belong to various families and exhibit different strategies for obtaining nutrients from host plants.

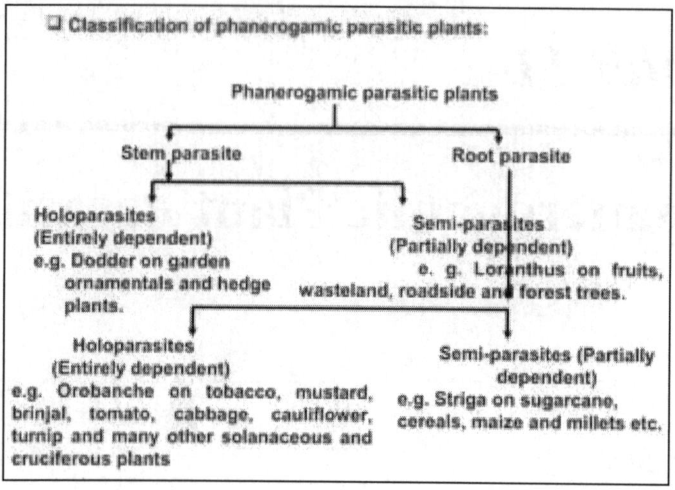

Here are a few types of phanerogamic parasites:

Obligate Root Parasites

Broomrapes (*Orobanche* spp.): Broomrapes are root parasites that lack chlorophyll. They attach to the roots of host plants and derive nutrients from them. They are known to parasitize a wide range of host species, especially in the sunflower family (Asteraceae).

Stem Parasites

Dodders (*Cuscuta* spp.): Dodders are stem parasites that wrap around the stems of host plants and form haustoria to penetrate and extract nutrients. They are adaptable and can parasitize a variety of host plants.

Aerial Parasites

Mistletoes (*Viscum album, Phoradendron* spp.): Mistletoes are aerial parasites that grow on the branches of host trees. They have specialized structures called haustoria that penetrate into the host's vascular system for nutrient absorption. Mistletoes are known for their association with certain tree species.

Holoparasites

Sandalwoods (*Osyris* spp.): Sandalwoods are holoparasitic plants that completely lack chlorophyll. They parasitize the roots of other plants and obtain all their nutrients from the host. They are found in arid and semi-arid regions.

Hemiparasites

Indian Paintbrush (*Castilleja* spp.): Indian paintbrush is an example of a hemiparasitic plant. While it can photosynthesize to some extent, it also forms haustoria to connect to the roots of nearby host plants, extracting water and nutrients.

Partial Stem Parasites

Yellow Rattle (*Rhinanthus* spp.): Yellow rattle is a partial stem parasite that parasitizes grasses. It forms haustoria on the roots of grasses and reduces their growth, making it a component of some grassland ecosystems.

These examples showcase the diversity of phanerogamic parasites in terms of their host specificity, habitat preferences, and adaptation strategies for nutrient acquisition. Studying these plants provides insights into the complex relationships between parasitic plants and their hosts in different ecosystems.

Dodder (Love vine, Amarbel)

Dodder, also known as love vine or Amarbel, is a fascinating parasitic plant belonging to the genus Cuscuta. Here are some key characteristics and information about Dodder:

- **Lack of Chlorophyll:** Dodder is unique among plants in that it lacks chlorophyll, the green pigment essential for photosynthesis. As a result, Dodder cannot produce its own food through photosynthesis.
- **Parasitic Lifestyle:** Dodder is an obligate parasitic plant, meaning it relies entirely on other plants for nutrients. It establishes connections with host plants to extract water, minerals, and organic compounds required for its growth and development.
- **Host Recognition:** Dodder is known for its remarkable ability to recognize and locate suitable host plants. It uses chemical and tactile cues to find potential hosts. Once a host is identified, Dodder extends its specialized structures, called haustoria, to penetrate the host's vascular system.
- **Appearance:** The appearance of Dodder varies, but it often forms thin, twining stems that wrap around the stems or branches of host plants. The color of Dodder can range from yellow to orange, giving it a distinctive and somewhat ethereal appearance.
- **Life Cycle:** Dodder begins its life cycle as a seed. When the seed germinates, the young Dodder plant seeks out a host plant. Once in contact with a suitable host, Dodder forms haustoria to establish a connection and begin extracting nutrients. As Dodder matures, it continues to spread and twine around its host.

- ☆ **Ecological Impact:** While Dodder is intriguing from a biological standpoint, it can pose challenges in agriculture. It has the potential to parasitize crops, leading to reduced yields. Controlling Dodder in agricultural settings is often challenging due to its ability to adapt to various host plants.
- ☆ **Economic Importance:** Some species of Dodder have economic importance in traditional medicine and are used in herbal remedies. However, its parasitic nature also makes it a concern for farmers, particularly in regions where Dodder-infested crops can suffer from reduced productivity.

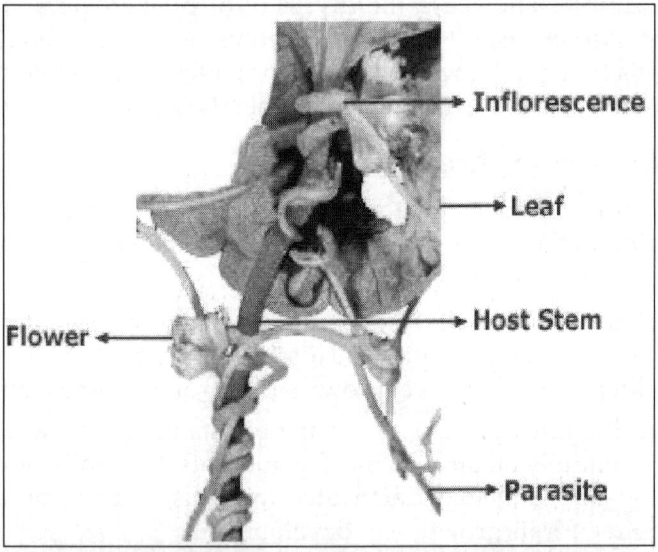

Understanding Dodder's biology and interactions with host plants is essential for developing strategies to manage its impact on agriculture and ecosystems. Researchers continue to study Dodder to unravel the molecular and physiological mechanisms underlying its unique parasitic lifestyle.

When stem of parasitic plant comes in contact with host, theminute root like organs. *i.e.* penetrates into the host and absorbs.

Mechanism

- ☆ The haustoria penetrate the stem or leaf and reach into the cortical region just outside the pericycle like an adventitious root.
- ☆ The haustoria secrete enzymes into the plant tissues that hydrolyze reserved food stuffs, such as, starch, and thus make them available to the dodder plant.

☆ These hydrolyzed substances and water are absorbed by the haustoria and are transported to the dodder stem where they are utilized for further growth and reproduction.

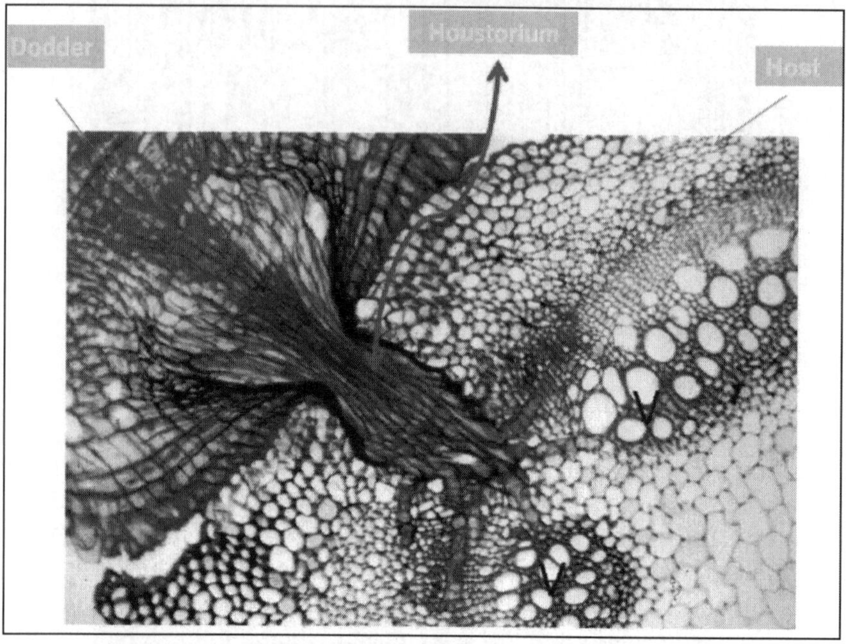

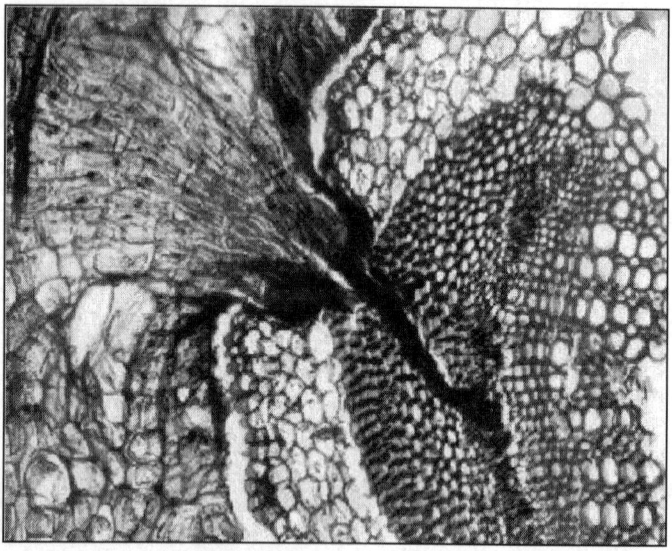

Overwintering and Dissemination

☆ As impurity of the crop seed.

- ☆ As seeds and stem pieces moved b yirrigation water.
- ☆ As stem pieces present on the dry straw from infested fields.
- ☆ As seeds in the manure.
- ☆ As stem pieces transported by cattle and farm implements.
- ☆ As stem pieces carried by birds or strongwinds.

Symptoms on Host Plants

- ☆ Host suffers from malnutrition
- ☆ Vitality of the host plant is reduced
- ☆ Plant may be dwarf
- ☆ Yellowing of the leaves with less flowers and fruiting
- ☆ In case of severe attack, the whole plant of the affected part may die.

Control Measures

Prevention

- ☆ Use of dodder free seed
- ☆ Cleaning agricultural implements
- ☆ Limiting movement of domestic animals
- ☆ Using of dodder free manure
- ☆ Restricting the flow of irrigation water

- ☆ Following a five year crop rotation with a non-host crop
- ☆ Avoiding the dodder infested cattle fodder.
- ☆ Destruction of the crop by burning before seeding.

Control

- ☆ Chemical control by spraying herbicide like chemicals *e.g.* 2,4-D, PCP (pentachlorophenol) @ 0.3-0.4 per cent
- ☆ Cutting or burning of patches, will kill both dodder and host plants but prevent dodder from spreading and from producing seed.

Systematic Position

Class: Dicotyledonae

Order: Solanales

Family: Cuscutaceae

Genus: Cuscuta

Species : Cuscuta gronovii

Mistletoes/Banda

Scientific name: Lorunthussp/Dendrophthae sp.

Semi-parasitic plant of fruit and forest tree*e.g.* Mango

Dense cluster of twigs

Smooth green leaves

Overwintering and Dissemination

- ☆ Overwinters as seeds.
- ☆ Spread by dispersal of its seed mostly through birds and to some extent by other animals.
- ☆ Birds are attracted by brilliant color of the fruit.
- ☆ Droppings of birds containing seeds also help in dissemination of the parasite

Chapter 12

Management of Plant Diseases

Integrated Disease Management

The intensive cultivation of genetically uniform and high yielding varieties of various crops coupled with excessive fertilization, close planting and frequent irrigation have contributed to aggravation of plant pathogens. The attention is focus on a system approached to keep the pathogens at a sub economic level with in population. A lot of information on this aspect is available which indicates that various approaches can be employed in judicious manner for integrated disease management

Management

1. Soil management
2. Seed Management
3. Environmental management
4. Irrigation management
5. Crop Management

Main Component of Integrated Disease Management

a) Prevention
b) Monitoring
c) Disease diagnosis
d) Knowledge of Acceptable disease threshold
e) Selection of management tools

Crop Management

a. Physical management
b. Cultural Management
c. Biological management
d. Chemical management

Crop rotation is single management practice, which is more effective in the elimination of soil borne pathogens. A micro culture of crops increases the population of microorganisms and permits them to alive on and within the root of those crops. In contrast, crop rotation help prevent their build up, as pathogens of one crop are not likely to attack another unrelated crop. Adjustment of date of sowing is may be profitable practices to circumvent attack of disease by avoiding the peak period of attack. To put this concept into practice the knowledge of the conditions favouring appearance of disease and vulnerable age of the plant at which they are susceptible to attack is essential. The heat therapy is effective in case of internally seed borne diseases where the mycelium remains in the embryo in the dormant form. Hot water treatment eliminates mycelium of smut diseases. Mulching is used to reduce the over summering inoculums of the pathogen responsible for soil infestation. Plastic mulching is reduced the inoculums potential of *Pythium, Taphrina* and *Protomyces.* Excessive use of water and high doses of nitrogen in general prolong the period of crop which may be attacked by diseases. Nitrogenous fertilizer tends to increase susceptibility of rusts, mildews and blights. Genetic uniformity of varieties has been recognized by several workers to be crucial factors in facilitating disease epidemics caused by air borne pathogens like rust of pea. Genetic vulnerability can be avoided by deliberate creation of variation in the host population. The cultivation of diverse host is an established effective method in plant disease management. Development and deployment of cultivars with durable resistance to plant diseases is a very economic and ease methods of plant disease management. Tools of molecular biology are now being used to facilitate the conventional disease resistance breeding programme and to shorten the duration required to develop a resistant cultivars in different crops. Molecular markers closely linked to resistant genes are used to identify the possible interogression of R-genes in susceptible cultivars upon hybridization. Selection of a particular trait in segregation population by the use of molecular markers is termed as Marker Assisted Selection (MAS).Use of DNA markers in plant disease resistance breeding programme have considerable increased in the country. There is a huge data on the molecular tagging of disease resistance gene irrespective of crop.

One of the common means of controlling plant pathogens in the field, green house and storage through use of chemicals that is toxic to the pathogens. Such chemicals are broad spectrum *i.e.* they are toxic to all or most of the pathogens. The efficacies of fungicides have been tested against pathogens in different crops. A disease prediction models is of fundamental importance for successful and efficient use of suitable chemicals for the managements of diseases. The feasibility of fungicide control has been restricted to economically protected situations and is not considered safe due to adverse effect on ecosystem. Hence, there is enthusiasm of searching mycoparasites to exploit tham as biological control agents. Antagonistic action of *Chaetomium globosum, Trichoderma* spp, *Gliocladium* spp., *Pseudomonas* spp., *Bacillus* spp. *Streptomyces* sp. *Penicillium* spp. and *Aspergillus* spp. has been found to be effective against many plant pathogens. Kalisena a bioformulation from strain of *Aspergillus niger* (AN-27) was found effective against many soil borne pathogens. Several phylloplane bacteria (belonging to Pseudomonas, Flavobacteria, Aeromonas) were used successfully as bio control agents for bacterial blight. Eco-friendly neem based products were effective as well as compatibility with chemical control. A few bacteriocin and HCN producing fluorescence pseudomonad were developed. The bacterisation of seed not only improved seed germinability but also reduced the root and seedling diseases of several crops

Bio-pesticides

Agriculture is the main source of income of the rural masses in India and the disease accounts for more than 30 per cent losses every year. A number of microbes were found to be effective bio control agents for the disease

Origin of Bio-pesticides

As early as in 1920, it was notices that pathogens are suppressed if some antibiotic producing microbes are added to the soil. However, the actual development began in 1963, when America organized an International Symposium on "Ecology of soil borne plant pathogens-Prelude to biological control"The proceedings was editated by Baker and Synder (1965) The first standing committee on biological control was establish within American phytopathological society in 1976. The Beltsville Agricultural Research Centre Symposium in USA was land mark where plant pathologist, entomologist weed scientists and nematologist were brought together for the first time. It is well known that the pesticides results in environmental contaminations as they remain over there for a longer periods. Therefore the government is attempting to increase the use of bio-pesticides using botanicals and microbial formulations. Farmers are also willing to adopt these pesticides with an intention to keep them safe and generate higher yields According to estimate only 5000 villagers comprising of 0.1 million farmers have reaped the knowledge and benefits occurring from the use of bio pesticides. Productions of organic Vegetables are coming in India scenario

Bacterial Antagonists

The widespread commercial success in the control of the crown gall disease by an avirulent strain of *Agrobacterium radiobacter* K-84 has triggered the attention on the use of bacteria as biocontrol agent against plant pathogenic fungi.

Antagonism is the phenomenon involved in the interaction. Antagonism can be described as killing, injury or inhibition of the growth of one microorganism by another. The possible modes of bacterial action in the inhibition of microorganisms are 1) direct penetration 2) antibiotic production 3) completion for nutrient 4) stimulation of host defence and 5) antibiosis. Bacillus species are the promising bio control agents as they produce endospores that are tolerant to heat and desiccation. Seed bacterization with *Baccilus subtilis* has successfully controlled the disease of groundnut and onion *Fusarium* wilt and *Pythium* sp are controlled by the use of cell wall degrading lytic rhizosphere bacteria. It is also controlled by chitionlytic arthobacter species isolated from rhizosphere. Root diseases of crop plants are controlled by *Baccilus subtilis* with seed bacterization. Sneh *et al.* evaluated over 700 bacteria and actinomycetes isolated from *Fusarium* suppressive soil for their ability to induce suppressiveness to *F.oxysporium f.sp. cucumerinum*. Among non-fluorescent bacteria only *Enterobacter cloacae* inhibited chlamydospore germination of *Fusarium*. They pointed out direct correlation between siderophore producing bacteria and inhibition of chlamydospore germination in soil. The *B. megaterium* B153-2-2 was a potential candidate for the management of microflora in rhizosphere. Bacterization of carnation roots with WCS417r controlled the *Fusarium* wilt by inducing phytoalexin accumulation.

Pseudomonas fluorescens

(Bacterial Bio agent -Multiplied in liquid culture)

Bacillus subtilis

(Bacterial Bio agent -Multiplied in liquid culture)

Fungal Antagonist

Various fungi are involved in the bio control of soil borne diseases as they exhibit widespread mycoparasitism and antagonism towards pathogenic forms. Fungal bicontrol agents like species of *Trichoderma, Gliocladium, Penicillium* and *Aspergillus* are used in the bio control of root diseases several example are available for bio control of soil borne disease caused by *Fusarium, Sclerotium* and *Rhizoctonia* by species of *Trichoderma.Cucumber* wilt caused by *Fusarium cucumerinum* is controlled by introducing *T. Harzianum* and *P. flourescens.* Several other species of *Trichoderma* proved promising against F. moniliformef. sp licopersici in tomato. Antibiosis, competition and mycoparasitism are the phenomenon involved in the Trichoderma plant pathogen interactions.

Gliocladium catenulatum was found to be antagonistic against root pathogens by exerting similar antagonistic effect as that of *Trichoderma.* Tomato crown rot caused by *Fusariun oxysporum f. Sp. radicislycopersici* was reduced by *Penicillium funiculosum* along with other antagonists in the field. Cabbage yellows disease incited by *F. oxysporum* is suppressed by *F.oxalicum* and *Trichoderma viride.* Pathogen of Water melon and Muskmelon wilt was suppressed by *Aspergillus niger. Aspergillus niger* was also found effective against brinjal and linseed wilt. Pre and post damping of was controlled by *T.viride, Streptomyces griseus and B.subtilis.* Antagonistic fungi are exposed to the fungicide applied to the soil so the strain.

Trichoderma harzianum
(Multiplied in sterile coffee husk)

Soil Solarization

Soil solarization is a method of hydrothermal disinfestations accomplished by covering moist soil with transparent plastic sheet during summer. The process is simple non hazardous and does not involve toxic materials; it is appropriate use of solar energy for soil solarization to kill the target organisms. This technology can effectively employed for controlling of soil borne diseases in tropical countries like India. Direct thermal effects are the major phenomenon implicated in the control of soil borne pathogen during solar heating. Moisture plays a crucial role in reducing the population of pathogens thus soil irrigated before solarization gave higher lethal values. Soil solarization increased the antagonistic activity and inducing soil suppressiveness with increasing plant growth because major antagonistic microorganisms are agents of suppression to major and minor pathogens in soil through aggressive colonization of root and antibiotic production. Soil solarization is a disinfestations method, has potential advantages. It is a non-chemical method, non-hazardous to the user and does not involve toxic substances, to the host plant or to other organisms.

In this perspective, it is less expensive than other methods. Soil solarization technology can easily be transmitted by the farmers and can be applied in large areas manually and mechanically, used of this method has been reported to reduce the population of many soil borne pathogens including fungi, bacteria and nematodes as well as weeds. *Pythium aphanidermatum* was found favoured relatively high temperature of 35°C, which is approximately 5°C above that for intermediate *Pythium* spp. group. A clear and transparent polythene sheet used for soil solarization and plastic mulching may increase soil temperature upto 50°C, which killed the pathogens. Hence, soil solarization can helped for reduction of yield losses due to soil borne diseases.

Bio-fumigation

Bio-fumigation is used for the suppression of soil-borne pathogens and pests by the incorporation of *Brassica*, cabbage, cauliflower and radish leaves into the soil. The action of volatile compounds essentially isothiocyanates (ITC's) produced by the hydrolysis of glucosinolates which are found within *cruciferae*. Glucosinolates have little biological activity, however, products formed from their hydrolysis (isothiocynanates) are considered to be the most toxic. Suppression of some fungal pathogens by the ITC's liberated from *Brassica* plants has been shown to be superior that the synthetic fumigant metham sodium. Different crop of *cruciferece* are used as bio-fumigants because of their different concentrations and type of ITC's emission during their decomposition that are different in their toxicity against pathogenic fungi. The use of bio-fumigation with mustard induced high rate of rhizome germination along with reduced soft rot which has boosted the yield of ginger. These reductions were attributed due to production of volatile compounds from *Brassica* crops through a process known as bio-fumigation. It was observed that inoculums level of *P. aphanidermatum* reduced in bio-fumigated soil in comparison to untreated soil. Reduced level of damping-off disease was also found in greenhouse seedlings with bio-fumigation and enhanced crop growth such as stem height and stem diameter. Soil infested with *R. solani* have controlled, with the application of mustard seed meal and found that percentage of healthy plants was 20 per cent in comparison to 60 per cent pathogen-infested control treatment with seed meal mustard bio-fumigation. Organic amendments, particularly plant residues and animal manure, augment the biocidal activity of bio-fumigation through the production of biotoxic volatile compounds emanating from the decomposition of organic materials. The numbers of *P. ultimum* and *S. rolfsii* propagules were reduced by more than 95 per cent, when the exposed to volatile compounds from heated cabbage or compost amended soil for 14 days.

Botanical Pesticides

Green plants are known to provide a vast reservoir of potential drugs and not more than 10 percent of these have so for been examined in detail for their biological activity against human diseases and much less against plant diseases.

In history poisonous plants are mention those have pest control properties like Indian Rig Veda, The classic book of Hinduism, compiled during second millennium. Democritus tried plant extracts for controlling plant diseases as early as in 470 B.C. Decoction of tobacco and alder buds with lime and sulphur against mildew diseases of trees was recommended by Forsyth. Bradely used pepper dust to protect trees against blight to their blossom. Several plants are also known to possess antifungal properties against *F. udum*. Leaf extract of Neem (*Azadirachta indica*), Mehandi (*Lawsonia alba*), Onion (*Allium cepa*), Garlic (*Allium sativa*), Tulsi (*Ocimum sanctum*) and Eucalyptus (*E. globules*) has been found effective against wilt of pigeonpea caused by *F. udum*. The bioefficacy of phytoextracts, were tested *in vitro* against *F. udum* causing wilt of pigeon pea. The garlic bulbs extract was significantly superior to inhibit the growth of *F. udum*, which was followed by turmeric bulb and neem leaves extract tested by poisoned food technique Screened fresh leaf extract (5 per cent) of A. *indica*, *Dhatura fastousa* and *Tagetes erecta* which were found most effective against *F. udum*. *In vitro* efficacy of bitter guard, turmeric, garlic and black papper have showed considerable diminution in the growth of *Fusarium udum*. Growth of *Fusarium udum* has been reduced by 15 per cent concentration of turmeric (89.2 per cent) followed by garlic (88.20 per cent) black papper (82.22 per cent). The bitter guard extract is found least effective against *F. udum*. Mandhare and Suryawanshi have screened 35 botanicals and found that *Ocimum sanctum*, *Eucalyptus* spp. and *Nerium indicum* completely inhibited the growth of fungus on PDA as well as Potato Dextrose Broth. The growth of *F. udum* was inhibited by 80, 75 and 71 per cent with extract of *Azadirachta indica*, flax line and *Hibiscus rosasinensis*, respectively. The extract of *Vitex negundo*, *Sorghum bicolor*, *Piper nigrum* and *Zea maysdis* were also effective and inhibited the growth of *F. udum* by 60, 60, 60 and 62 per cent, respectively. Neem Kernel extract (*A. indica*) was effective in suppressing mycelium growth of *F. udum in vitro* and controlling wilt disease *in vivo*. Garlic bulb crude extracts was reported to inhibit the spore germination and mycelia growth of different species of *Fusarium* causing wilt diseases in vegetable and horticultural. Leaf extracts of *Citrus media*, a root extract of *Asparagus adscendens*, rhizome extract of *Curcuma longa* rhizome extract of Zingiber *officinale* and a bulb extract of *Allium sativa* and found that the higher concentration of plant extract completely inhibited (100 per cent) the mycelial growth of *F. udum* both *in vitro* and *in vivo*. *Azadirachta indica* leaf extracts of 100 per cent concentration completely inhibited spore germination of *Fusarium* spp. while 100 per cent *Allium sativum* leaf extract partially inhibited spores of *F. equiseti* (20 per cent germination). *Ocimum sanctum* leaf extract 100 per cent inhibited the growth of fungus because it have antifungal properties, against the spores of *F. solani*, *F. oxysporum* and *F. equiseti* as compared to control the inhibition was 90, 94 and 90 per cent respectively. Five medicinal plants (*Strychnos nux-vomica*, *Calotropis procera*, *Azadirachta indica*, *Ocimum sanctum* and *Allium sativum*) were evaluated on spore germination of different species of *Fusarium* and found complete inhibition by 100 per cent aqueous extract of

Azadirachta indica while only 20 per cent germination of *F. euiseti* spores was recorded with treatment of 100 per cent aqueous extract of *Allium sativum*

Equipment used in Plant Disease Control

Pesticides used in different forms *viz.* spray, dust, fog, mist and smoke. On the basis of pest, used of chemical and application machine are essential to choose the right type of machine to performed effective management.

Plant protection machines can be classified as follows:

- Sprayer
- Dusters
- Sprayers cum dusters
- Seed dressing machines
- Fumigators
- Flame throwers
- Bird Scorers
- Mechanical seed eliminators

Sprayers

A sprayer breaks up the liquid in to droplets of effective size and distributed them uniformly over the surface or space to be protected. Another function is to regulate the amount of insecticides to avoid an excessive application that might prove harmful or wasteful. Nozzles are the most important. These should be selected on their ability to deliver the desired amount of spray and deposit it uniformly on the surface which has been treated. The common types of nozzles used in sprayers are flat fan, hollow cone and solid cone. The flat fan nozzles are essentially low pressure and low volume in character. A high pressure in this nozzle produces too many small drops which drift away. The flat fan nozzle delivers more spray towards the ends than in the centre or vice-versa after some time. It is used for especially weed control work. The hollow cone nozzle is made to liquid in swirl in a chamber, The film of liquid is escapes through a single hole and as it rotates. The performance of nozzle cone is affected by changing the swirl plates and altering chamber depth. These nozzles work with wide pressure range and are suitable for medium and high volume spraying. It does not give as uniform spray in the field. It is used for spraying in vegetable and field crops. The solid cone nozzles has the same mechanism as hollow cone nozzle except swirl plate bears a central hole It allow a faster application rate upto 5 g.p.m. at high pressure but it can work at low volume nozzle at low pressures. This type of nozzle produce large drops which vary very much in their sizes. It is suitable for spraying high trees from the ground. High concentration of the insecticide especially when formulation of water dispersible powders are used, affects the droplet size. The more concentration of powder in water, the less is the atomization and larger is the production of droplets. When the spray liquid becomes sufficiently thick in concentration, it starts coming in the form of threads from the nozzle instead of droplets and at time clogs the nozzle. The water dispersible powders are not considered suitable for low volume spraying.

Type of Sprayer

Hand Sprayer or Atomizers

These are two types one is single action and other is continuous action. Both develop 0.9-2.25Kg of air pressure and have 1/16 inch to 1/32 inch diameter of nozzle orifices and produce droplets 15-50 micron in size. They are used to apply sediment free spray materials. The single action type of sprayer acts only on the stroke of the pump and is provided with non adjustable nozzle. It is used for spot disinfection

The continuous sprayers develops pressure after few strokes of the pump and gives continuous spray of uniform pattern, it have interchangeable and adjustable nozzles which are able to deliver fine to coarse droplet spray. These are suitable for spraying in small plants in kitchen gardens.

Compressed Air Sprayer

It has a tank to contain the liquid and air. Tank is fillied with liquid in ¾ capacities. In the liquid, the air pressure is created by means of a hand pump, which pressurizes the liquid from the top and forces it to go outsides in the form of spray. A working pressure of 22.5-31.5kg per square inch is created. It can discharge 45-91litre of spray liquid per hour. It is used for spraying field crops, vegetables and nursery plants by carried on the back or shoulder of the operator. The essential parts are a pump with handle, pressure gauge, an air release, valve, liquid filling port, air jet agitator and back rest, delivery hose pipe, and spray lance with nozzles.

Battery Knapsack Sprayer

They are similar to compressed air sprayers except that air and water is charged into them through a central charge pump. The charge pump is hand or power driven. It need not be charge again and again and can be used conveniently.

The hand charge pump works like as ordinary pump for filling in air in the sprayer but when connected with a suction tube, it sucks liquid and pushes it into the sprayer tank till desired pressure is obtained. It can control about 6 sprayers. Such sprayer worked at a total pressure of 31.5-45kg. /sq. inch. They are suitable for spraying all types of crops including tea and coffee but not grown up trees.

Knapsack Sprayer

This pump is operated constantly to get a continuous discharge of liquid. It is fit comfortable on the back of the operator. The lever handle provided at the bottom or on the top of the tank. A few stroke of the pump build up pressure in the air chamber located within the tank or outside. The handle is worked constantly so long sprayer is used. They build a pressures of 80-150 lbs with boom attachment it is possible to spray an area upto 2 acres/day with one sprayer.

Bucket Sprayer

It is designed to be used in a bucket containing spray liquid. It is provided with a pump which may be single or double acting. In latter case it discharges on both directions *i.e.* downwards and upwards strokes of handle Good sprayers can develop pressure upto 54kg/squire inch. It con used for small area in the garden or nursery beds.

Wheel Barrow Sprayer

It has a tank mounted on a single wheel or double wheel. The sprayer consist double acting hand pump fitted inside the tank. The capacity of the tank varies from 22.7- 68.1litre. It is operated by lever like handle and provides pressure inside the tank which can spray continuously without operating handles. The pressure develops by this sprayer ranges from 90 to 135 kg/ sq. inch. It is adequate for spraying orchards, medium sized fruit trees, nurseries, gardens and crops like cotton, sugarcane, potato in a day's 2-3 acres can be covered.

Slide Pump Sprayer

It is small hand operated sprayer comprising of telescopic plunger type pump. It have discharge pipe and suction hose with strainer. The sprayer is operated by back and forth motion of the plunger after dipping the suction hose in a bucket. It gives a continuous discharge of spray liquid and develops pressure upto 54kg/sq. inch (p.s.i.). It is suitable for ornamental pots, Kitchen garden and nursery plots.

Rocking Sprayer

It is provided with a pump and a pressure fitted on a wooden board and is operated by means of a handle connected with the pump. It may be double acting or single acting depending on the pump with one or two lances to spray. Rocking movement of handle helps in building up the pressure in the pressure chamber which forces the sucked in liquid into a spray through the nozzle. The part of sprayer is suction hose with strainer, a delivery hose with spraying lance. Before operating the sprayer, the suction hose is dipped in a liquid containing tank. It develops 90 to 112.5kg pressure and gives output of 68.1-90.8litre of liquid per hours with one nozzle. About 2 acres can be covered a day. It is suitable for medium size trees, crops and gardens.

Foot Sprayer

It is operated by foot with same principle as hand operated, in this sprayer constant pedaling is required after dipping the suction hose in a drum of bucket and holding the sprayer with one or both hands. The pump is fitted on an iron stand and a pedal is attached with plunger rod and operates the sprayer by its upward and downward movements. Foot sprayer develops 112.5-135kg pressure p.s.i. it has an output of 113.5-136.2 litre/hour with one nozzle. About 2-3 acres of medium sized crop can be easily covered per day.

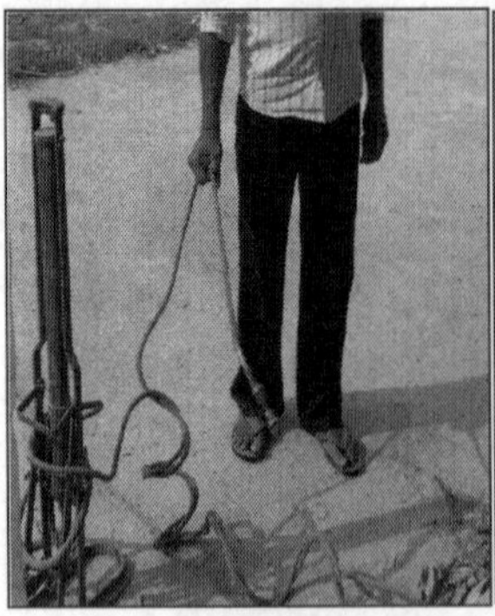 

Power Sprayer

These are run by petrol or kerosene oil or electricity and are mainly three types:

1. Hydraulic sprayers
2. Mist sprayers
3. Fog or smoke generators

Hydraulic Sprayers

In hydraulic type power sprayers, the pressure created by the pump is forced out through the nozzle the spray liquid is generated in spray fluid. It is forced out through nozzles in small droplets. The droplets in this case are deposited on the surface by sheet force of the spray and are mostly 150-300 microns in size. The engine is generally 1 h.p to 20 h.p depending on the pump capacity and pressure range. Pump developing pressure upto 800 lbs p.s.i. or more can also be used. The satisfactorily at lower pressure range of 22.5-45 kg p.s.i. The essential parts of a hydraulic sprayer are engine, a pump, a tank with agitator, framework with or without wheels for mounting the engine, pressure regulator gauge, control valves, action and delivery hoses, strainer and nozzles. Hydraulic sprayer can be grouped into the following main types-

Multipurpose Farm Sprayer

These sprayers can work at reduced pressures at 30-50 lbs p.s.i. at high pressure at 90-360kg p.s.i. When spray required on weed or crop and fruit trees respectively. The tank has a capacity of 227-908 litre or more and is provided with a mechanical agitator. For spraying it is provided with booms or spray 5-10 acres/day.

General Use Sprayers

Under this category fall the small wheel barrow or skid type of sprayers with discharge capacity of 6.81 to 13.62 litre/ min. at pressure upto 112.5kg. The power is furnished by on air cooled engine of 2/3 to 2 h.p and tank capacity varies from 12.5 to 18 gallons. They are provided with a mechanical or jet agitator, a pressure gauge, pressure regulator and a relief valve, spray gun, suction delivery hoses and handles for carrying it easily in the hand. A sprayer of this type can normally cover 3-5 acres/ day.

High Pressure and High Volume Sprayers

These types of sprayers are used for spraying toll trees and dense crops. They discharge 36.32-272.4litre/minutes at pressure ranging upto 450kg. The tank size is 1362- 2724litre. The field of potato, tomato, cabbage and cauliflowers are used for pressure spraying. In a day 10-20 acres or more can be covered

Low Pressure and Low Volume Sprayers

For spraying pressure lower than 45kg p.s.i. low volume spraying have been desired to meet the requirement. They are used to spray crops with a smaller quantity of spray liquid at low cost. Their use is limited in as much as only sediment free type materials can be sprayed. They can be mounted on tractor a wheel mounted trailer. Such sprayers are provided with booms. One sprayer covers approximate 15-20 acre/days.

Self Propelled High Clearance Sprayer

They are used for spraying tall ground crops which cannot be sprayed by other conventional types of sprayers. Plants size upto 5-8 fit in height are easily sprayed by booms. The booms are adjusted accordingly to the height of the crop and are fed by generally a low pressure, low volume sprayer mounted on the self propelled carrying frame or tractor. Such sprayers have 3-4 wheels and power is applied to the rear wheel when tractor is not used for moving them. About 10-15 acres are covered/ day.

Hydro Pneumatic Sprayer

There liquid is carried in a pressure tank and the spraying pressure is developed by means of an engine powered air compressor, thus avoiding the need to pass liquid to this pump. They are either skid type are mounted on wheels and work at maximum pressure of 33.75-45kg. In this sprayer discharge is denoted in cubic feet per minutes volume at the desired pressure equals 29.51 litre per minutes. It is not used for spraying in crops.

Mist Sprayers

They are essentially low volume sprayers developed to apply insecticides in a concentrated form and are known as air blast or drift sprayers. The dilution of spray materials is reduced by 20-80 per cent or more. There is saving of chemicals used due to the production of smaller droplets Chemical saving is gave due to the production of smaller droplets 50-150 µ, which is a run off the from foliage. These are used for treating large areas of fruit trees, high and dense crops *etc.* The mist sprayer is fitted with a low pressure low volume pump which forces the spray materials under low pressure. It emits one gallons liquid per 10minutes and cover 150 to 250 acres in an hours.

Fog/Smoke Generators

They disperse spray material in extremely fine droplets (1-50 µ) in diameter which remain air borne for a long time. It can be successfully used in dense foliage or closed space where fog take time to settle and does not drift with the wind. It has the advantage of a uniform coverage of the object with chemicals.

The term high, low or medium volume spraying signifies the rate of application of the spray material over a given area. The rate of application is related to the crops being treated. The low volume sprayer is used for spraying citrus and mango and high volume is used for ground crops such as ground nut, paddy, potato, brinjal *etc.* The spray material is deposited over the surface of the leaves either in large drops or in small droplets. In high volume spraying the liquid will not spread in the form at a film then accumulation at the edge and finally run off on the ground. In medium spraying, the liquid tends to deposit on the leaf either in large drops or in small pools where large drops coalesce without runoff. When the spray is deposited in small drops discretely which do not form pools it becomes low volume spraying.

High volume spraying- Over 272.4litres/ acre

Medium volume spraying- 90.8-272.4 litres / acre

Low volume spraying- Upto 90.8litres/ acre

The volumes are only guides. It is the coverage and spray deposit which are important for the control of insect pests or plant diseases. The bio efficiency of a deposit depends on:

a) The amount of active ingredient deposited
b) The distribution of active ingredient based on coverage of plant surface

The degree of coverage will depend on the nature of the organism to be controlled and the mode of action of the toxicant. Because highly mobile insect are need to cover thoroughly with chemicals. The main problems of high volume spraying is the obtaining and carting of huge quantities of water which is time consuming and add to the cost of spraying. Besides the output of work is less and the initial cost of high volume sprayer is more.

The principal advantage of low volume spraying lies in the saving to labour, active ingredient and time. In low volume spraying machines cast is comparatively less and output of work is high. Pesticides formulations are very important. The best type of formulations for use in low volume sprayers is the miscible liquid which could spray without sedimentation in the tank or blocking the nozzles for ting jets. They are three types-

- ☆ Air blast spraying 227-2270litres/tank
- ☆ Mist sprayers/blowers 45.4-90.8litres/ acre
- ☆ Drift sprayers 9.8-22.7 litres/acre

Advantages of Spraying

- ☆ Greater range of formulation is available
- ☆ Better utilization of chemicals
- ☆ Rate of application per unit area can be increase or decrease with droplets size
- ☆ Uniform deposition and better adherence
- ☆ Cost of spraying is less

Disadvantages of Spraying

- ☆ Less output of work per unit area
- ☆ Equipment is generally heavier
- ☆ Preparation of pesticide formulation is difficult
- ☆ Spray formulation is requires carefully storage
- ☆ Application cost is high

Dusters

Pesticides may be applied in dust formulations with specially designed machines called dusters, which are mainly hand driven and sometimes power driven. All dusters essentially consist of a container which is termed as a hopper of a system of keeping the dust particles agitated and in motion and to be fed into a current of air so that the particles are ultimately discharged in the form of fine dust.

Type of Dusters

1. Hand Dusters

- ☆ Package or container dusters
- ☆ Hand pumps
- ☆ Bellown type dusters
- ☆ Rotary hand dusters

2. Power Dusters

- ☆ Tracion duster
- ☆ .Engine operated power duster

Container Dusters

These are primitive. The dust is kept inside a plastic rubber of leather containers. By squeezing or through a crude valve system dust is discharged into a small stream. These are useful for kitchen garden and for domestic use.

Hand Pumps

They are small and operated by a plunger pump. The container is normally straight tube with nozzle. Dust is thrown out through the outlet by the compression stroke of plunger pump. These dusters have very limited use in kitchen gardens, houses or on animals. Plunger dusters are simple in design and consist of an air pump.,

Bellow Type

They are operated by the expansion and contraction of a pair of bellows during which process dust in sucked in and then thrown out into the delivery system. A more convenient type of bellow duster is designed in the knapsack fashion. The hopper is of longer capacity.

Rotary Hand Dusters

It is also known as Crank type duster which is extensible used for dust application over large area of crops. These duster operated by the current at air produced by a fan which is driven by the rotary motion of a handle through a reduction gear. The fans draws dust from the dust chamber and discharge it into the delivery pipe from where it is blown out through one row or two row nozzles. Normally 0.5 to 150 gm/ minutes may be discharged at 35 rpm. 2 to 3 acres can be covered with rotary duster in a day.

Fan Type Hand Duster

Power Dusters

The basic components of a power duster are essentially the same as in a hand duster. The power for operation is generated from the wheels, engine or vehicles on which they are mounted. The power dusting machine comprises a hopper containing one or more agitators, a regulator for discharging of dust, a fan or blower producing air blast for drawing the dust from hopper and blowing its outside. The size of hopper varies with capacity to holding 60 to 200 pounds of dust depending on the density.

Wheel-barrow Dusters

These types of dusters are provided one or two wheels and drawn by hands or animals. Such dusters have greater hopper capacity ranging from 9-22.5kg and can be used for dusting row crops in larger area, where crank or power dusters are not economical.

Tractor Mounted Dusters

The tractor mounted dusters deriving power from the tractor are provided with power from the tractor are provided with power take off sleeve coupling or a belt and pulley for mounting directly on the power take- off shaft. No separate power source is needed. The capacity of the hopper varies from 20-45kg.

Engine Operated Dusters

Normally 1-3 h.p. air cooled engines are used and the hopper capacity is usually 10-20 Kg. Power duster may have a single outlet or may be fitted a series of 4-8 outlets with flexible pipes and fitted with a boom with as many as 18 nozzles. Dust can be discharged at the rate of 1-9 Kg/min. A power duster with one out let may cover 12ha/hr. dust flame may be regulated from 0-1.5 kg/min by adjustment at multi- hole discs. Large sized dusters with hopper capacity of 50-100 Kg are operated by engines upto 25 hp with the introduction of low volume spraying. The concentrate spraying materials of dusting is no longer palpated to the extent as it was before.

Knapsack Type Spray Dusters

This type of duster is operated by 1 h.p. to 2 h.p. engines and can be carried on the back of the operator. The hopper and the delivery hose pipe in such types of machines are small and feeding capacity is 4.5-5.4kg of dust and dust discharge is about 1 to 4 pound per minutes. Such type of duster is operated by one man. A spray dust can be done 10 acres per day

Advantage of Duster

1. Less deterioration of dust formulation
2. Dusting can be done in high wind velocity
3. Dusting equipment is simpler

4. Less cost of treatment
5. High work can be done per unit time

Disadvantage of Dusters

1. Higher cost of active ingredient
2. Un even deposition and poor adhesion
3. Range of particle size is fixed
4. High transportation cost due to bulk of dust
5. Dusting can be done only morning hours

Seed Dressors

Many type of seed dressers are available and they vary in capacity

Type of Seed Dressers

1. Rotary seed dresser
2. Cascade or gravity seed dresser
3. Slurry seed dresser
4. Grain treating machine
5. Sack impregnating machine

Rotary Seed Dresser

It consists of a drum, fitted on a stand which can be rotated with the help of a handle provided at one end. The drum has opening to introduce seed and chemical, which take out seed after treatment. The drum should be rotated slowly 30-40 times during the process seeds will be thoroughly coated with the chemical.

Gravity Seed Dresser

The machine has a steel drum fitted with cones and baffles which are perforated. A lid is provided at the top through which the seed and chemicals are introduced. Seeds while coming downwards slowly due to gravity get mixed with the chemical. This is used for chemicals which are comparatively volatile in nature. It take more time but does not involve any manual labour.

Slurry Seed Treatments

These machines are driven by 3-4 hp electric motor or oil engine measured quantities of seed and chemical (slurry) are automatically fed into the treatment chamber in a synchronized manner. The slurry cap introduces a given amount with each seed lot into the mixing chamber. The seed is thoroughly mixed in the chamber to prevent sedimentation. Approximately 30 quintals of seed may be treated per hours in this machine.

Grain Treating Machine

This is designed by PAU Ludhiana. It has a hopper, grain metering device and a spraying unit. The spraying unit has a low volume nozzle; the spraying liquid is discharged at the rate of 290 ml/ min. at a pressure of 2.1 Kg/cm2. As grains which are fed on the top fall slowly due to gravity. They are treated with the chemicals these machines can treat upto 100 quintals of grains in 2 hours.

For Pulse

The pulse grains from the hoppers are moved through corrugated rollers into collecting container through a trough, while the grains are treated with chemicals, which is applied by a hand compression sprayer. About 100Kg of grain are fed in a period of 8-10 minutes.

Sack Impregnating Machine

This method has been developed by the CFTRI, India. Here the gunny bugs meant for storage of grains are treated. The sacks move through an endless they are impregnated with the chemical which is applied from a compressor in a 10 liter tank maintained at a constant pressure. The liquid is applied through low volume nozzles only one side is treated at a time approximately 3000 sacks can be treated/ day.

Soil Injectors

Types

- ☆ D.D.gun
- ☆ Tractor mounted applicators
- ☆ Granule applicator
- ☆ Aerosol projectors

D.D. Gun

It consists of a pump a container for the liquid to be injected and a Holland injection needle. The nozzles can be driven up to a depth of 25-30 cm. The pump then pushed down to force liquid into the soil containers have a capacity of 2.2 to 3.5 liters.

It is used for kitchen garden or experimental purposes.

Tractor Mounted Applicators

Field applicators consist of a container from which large fines come out. They are drowning like a harrow. The liquid is fed into the fines as the fines move through the soil.

Granule Applicator

The essential component are the hopper in which the granules are kept, a long flexible discharged tube with nozzle at the distal end and a finger controlled mechanism for the regulation of consistent flow of granules. They are usually made of plashes or light and have a granules capacity of 1.0 to 1.3 Kg.

Dust Applicator, Cymag Foot Pump

It is essentially a plunger type at duster with a dust chamber of 450g capacity, pump, and discharge pipe and delivery tube. The air blast agitates the powder and forces the same to be discharged through the delivery tube.

Aerosols Projector

These are useful in large green houses or in enclosed spaces or in go downs or store houses. Aerosol formulation is poured into the pressure chamber. Gas is liberated which exert pressure on the aerosol liquid on release of pressure by trigger. The liquid is forced through the nozzles overhear. The liquid on reaching outside because of the low boiling point vaporizes immediately with the formulation of fine mist.

Methods of Disease Management

Seed Treatments

Mechanical Separation: The seed lots are cleaned mechanically by separating the soil clods, plant debriah, weeds, discoloured, deformed and shriveled seeds showing visual symptoms of the disease by hand picking under stereobinocular microscope. The vegetative propagules are cleaned by excising the infected portions.

Hot water treatments: This technique is used for controlling various seed-borne pathogens. In this treatment, the seed with or without pre-soaking, is held in water at certain temperature generally close to the maximal temperature critical for both pathogen and host for a definit periods of time. Wet heat appears to be an effective agent for directly killing the pathogen.

Dry heat treatment: Dry heat treatment has been used by some workers occasionally. Different combinations of temperatures have been tried which given promising results against certain pathogens.

Solar heat treatment: This treatment is applicable in countries with a hot climate and plenty of sunshine. Grain is pre-soaked in water for some hours under shade or in room and then dried on ground in these layers under bright sunlight, usually at noon.

Vapour heat treatment: The dry seed are treated in vapour condition generally by pressure flow. It has also been used to treat the vegetative propagules for eradication of virus infection.

Chemical Treatment

Chemicals are used for controlling seed borne pathogens. Various fungicides are available which can be used as seed dressing or dip for the vegetative propagules. The chemicals also give protection against soil borne fungi. Systemic fungicides are widely used at present. The metalazyl seed treatment has been effective against the powdery mildew. Oxathiins Plantvax and vitavax are used against smuts. Benomyl compound are broad spectrum and are effectives against many seed borne fungi. Fungicides can applied with following methods-

Steep treatment: The seed is soaked in a fungicidal liquid solution or suspension for a definite period of time and fungicides solution is then drained and seed is dried.

Sprinkling treatment: The seed is sprinkled with a fungicidal liquid, solution or suspension, left damp with this treatment for a definite period of time and then dried.

Dust treatment: The seed is mixed with a fungicides dust in a mixing device until the seed is well coated usually @ 2 g/ Kg seed.

Slurry treatment: The seed is mixed with a dust fungicide in a special treater in which small calibrater amount of liquid 5-10 ml/Kg seed is added, thus forming a soup like slurry to ensure coating without undue wetting.

Wet treatment: The seed is mixed with relatively small amount of concentrated liquid about 100-300 ml/Kg seed. There is no recovery of liquid after treatment but drying is necessary.

Fumigation: The seed is treated in an airtight container or chamber for a definite period of time using volatile materials. A fungicides, nematicides or insecticides.

Pelleting: Application of sticker is necessary prior to mixing the seed with a dust which may contain pesticides or other materials. The seed is first coated with methocel. A cellulose acetate sticker in dilute solution and is then agitated with the dust in such a way that each seed becomes a pellets containing an outer zone of protectant.

Sprays and Dusts

Chemicals applied as sprays or dusts on the plants are usually aimed for control to pathogens. Most fungicides and bactericides are protectants and present on the surface of the plant in advance of pathogen, that can prevent to the entry of pathogens. Because their present on the surface of plant does not allow pathogens spores to germinate and they may kill spores. Some fungicides have direct effect on pathogens which have already invaded the leaves, stem, root and fruit in this situation fungicides act as eradicant and killing the pathogens inside the host or they may suppress the sporulations and multiplication. Some fungicides are partially systemic and some are clearly systemic, partially systemic fungicides are absorbed by aparts of plant leaf tissues and translocated internally into the whole leaf area example dodine.

Systemic fungicides translocated internally throughout the host plant and kill the pathogens *e.g.* benzimidazoles, benomyl, thiobendazole, oxanthiins, carboxin and oxycarboxin.

Spray of fungicides is more effectives than dusting of fungicides. Dusting is preferable during rainy weather because of adhere better to wet plant tissues.

Drenching

The oldest and best method of chemical treatments is the drench especially formaldehyde drench. 4.54 litre of formaldehyde (37-40 per cent) is added to 227 litre of water or in case of heavy disease infestations the strength may be ncreased to 4.54 litre of formaldehyde to 113 litre of water. The soil must be drenched with this solution to a depth of 3-6 inches. This require from 115-130 litre of solution per square feet of surface. The soil surface should be covered with polythin sheet for a 48 to 72 hrs after treatment. The soil should be forked or spaded before treatment. After treatment the soil must be aired until all traces of formaldehyde odor have disappeared. It required 10-15 days. The present of small amount of chemical in the soil will kill the young seedlings. This method is effective for soil borne disease management.

Fumigation

Fumigation is especially valuable as a method of controlling disease, nematode and pest in soil, greenhouse, warehouse, mills *etc.* where spraying is not possible and controlling insect, nematodes and diseases. The chemicals used for this purpose are usually volatile and on coming in contact with soil moisture release gases which diffuse in the soil and kill the larvae of the nematode and insect and spore of fungi. Application of these high volatile toxic substances is recommended few days advance sowing and planting of crops. Fumigating equipment is required for injecting the chemicals in soil. The diffusion of chemical in soil is restricted at regular distance 12-15 inches. The depth of application is kept at 6-9 inches. The effectiveness is depending on soil texture depth, soil temperature, moisture and pulverization.

Paste and paint: Wounds created during houing, pruning and training of trees often serve as openings for entry of pathogens. The protection of these wounds required protective layers of past and paint. These pastes are prepared with fungicidal chemicals such as copper sulphate and lime, copper corbonate, red lead and antibiotics *etc.* in suitable carriers. The past heel cut surface of wound for long enough and protect the entry of pathogens.

Basic Precaution in Pesticides Usages

A. Purchase

- ✰ Purchase only just required quantity e.g. 100,250,500 or 1000g/l for single application in specified area.

Identification of Pesticides Hazardous and Symptoms of Poisoning

Sl. No.	Colour of Toxicity Triangle in Pesticides	Classification as per Pesticides Rule, 1971	WHO Classification by Hazard	Symptoms of Poisoning
1	Yellow	Highly toxic	Class-II, Moderate hazardous	Nausea, vomiting, respiratory failure, coma, weakness, anxiety and death
2	Red	Extremely toxic	Class-1b, highly hazardous	Constriction of pupils, salivation, profuse sweating,vomiting diarrhea, epigastric, tightness in chest
3	Green	Slightly toxic	Table-5 (unlikely to present acute hazard in normal use)	Headache, palpitation, vomiting, irritation of nose, throat, eyes and skin
4	Blue	Moderate toxic	Class-II slightly hazardous	Nausea, vomiting, irritation

- ☆ Do not purchase leaking containers, loose, unsealed or torn bags
- ☆ Do not purchase pesticides without proper/ approved labels

B. Storage

- ☆ Avoid storage of pesticides in house premises
- ☆ Keep only in original container with intact seal
- ☆ Do not transfer pesticides in other containers
- ☆ Never keep them together with food feed/fodder
- ☆ Keep away from reach of children and livestock
- ☆ Do not expose to sun-light or rain water
- ☆ Do not store weedicides along with other pesticides

C. Handling

- ☆ Never carry/ transport pesticides along with food materials
- ☆ Avoid carrying bulk-pesticides on head, shoulders or on the back

D. Precautions for Preparation of Spray Solution

- ☆ Use clean water
- ☆ Always protect your nose, eyes, mouth, ears and hands
- ☆ Use hand gloves, face mask and cover your hands with cap
- ☆ Use polythene bags as hand gloves, hand kerchiefs ar piece of clean cloth as mask and a cap or towel to cover the head

- ☆ Read the level on the container before preparing spray solution
- ☆ Prepare solution as per requirement
- ☆ Do not mix granules with water
- ☆ Concentrated pesticides do not fall on hands etc. while opening sealed containers.
- ☆ Do not smell the pesticides
- ☆ Do not spilling of pesticides solution while filling the spray solution tank
- ☆ Do not eat, drink smoke or chew while preparing solution
- ☆ The operator should protect his bare feet and hands with polythin bags

E. Equipment

- ☆ Select right kind of equipment
- ☆ Do not use leaking, defective equipment
- ☆ Select right kind of nozzle
- ☆ Do not blow/ clean clogged- nozzle with mouth
- ☆ Use old tooth brush tied with the sprayer and clean with water
- ☆ Do not use same sprayer for weedicides and pesticides

F. Precaution for Applying Pesticides

- ☆ Apply only recommended dose of pesticides
- ☆ Do not apply on hot sunny day or strong windy condition
- ☆ Do not apply just before the rains and also after the rains
- ☆ Do not apply against the wind direction
- ☆ Emulsifiable concentrate formulations should not be used for spraying with battery operated ULV sprayer
- ☆ Wash the sprayer and buckets *etc.* with soap water after spraying
- ☆ Containers, buckets *etc.* used for mixing pesticides should not be used for domestic purposes
- ☆ Avoid entry of animals and workers in the fields immediately after the spraying

G. Disposal

- ☆ Left over spray solution should not be drained in ponds or water lines etc. Throw it in barren isolated area if possible
- ☆ The used/ empty containers should be crushed with a stone/stick and buried deep into soil away from water source
- ☆ Never re-use empty pesticides containers for any purpose

H. Warnings

- ☆ Consult expert of the field of plant protection before use of any pesticides regarding doses and time of application
- ☆ Do not used pesticides in wrong way
- ☆ Select pesticides wise fully
- ☆ Use pesticides judiciously on need based manner
- ☆ Do not use green leaves as food after spray of pesticides

Method of Pesticides and Fungicides Calculation

General Formula

$$F = \frac{Sc}{ai}$$

where

F = is the quantity of formulation of fungicides or pesticides is required

S = is quantity of spray fluid required

C = is the percent concentration of the fungicides or pesticides in the spray fluid

a.i = is the percent concentration of active ingredient in the formulated fungicides or pesticides

Example No. 1: How much spray fluid of 0.1 per cent concentration can be prepared from 125g of Bavistin 50WP

F = 125g

S = ?

C = 0.1 per cent

Ai = 50

$$F = \frac{Sc}{ai}$$

$$125 = \frac{S \times 0.1}{50}$$

$$S = \frac{125 \times 50}{0}.1$$

= 62500ml or 62.5litre

Protocols of Seed Treatment of Vegetable Crops through Hot-Water

Crops	*Temperature (° C)*	*Time (Minutes)*
Brussels sprouts	50	25
Broccoli	50	20
Cabbage	50	25
Carrot	50	20
Cauliflower	50	20
Celeriac	48	30
Celery	48	30
Chines cabbage	50	20
Coriander	52.5	30
Cucumber	50	20
Eggplant	50	30
Kohlrabi	50	20
Lettuce	48	30
Mint	47	10
Mustard	50	15
Spinach	50	25
Onion	46	60
Radish	52	30
Sweetpotato (roots)	46	65
Tomato	50	25
Turnip	50	20
Yam(tuber)	46	30
Okra	60	30
Tea	50	20
Sugarbeet	55	30
Pepper	10	10
Watermelon	51	10
Pumpkin	50	10
Turmeric	50	30
Guar	50	10
Bean	48	25
Betlevine	51	10
Pea	55	15
Potato	50	20
Cowpea	52	15
Ginger	50	30

List of Biological Control Agent

Name of Parasites/Predator	*Name of Parasites/Predator*
Trichogramma sp.	*Trichogrammatoidae armigera*
Compoletis choridae	*Carcelia illota*
Eriborus argenteopilosus	Hyposoter didynotor
Palexorista laxa	*Gonoipthalmus halli*
Telenomus remus	*Chelomus helioptae*
Cotesia colemani	*Microplitis* sp.
Charops obtuse	*Euplectus* sp.
Peribaea orbata	*Phaneraotoma* sp.
Chrysonomyia appannai	Coccinellids
Syrphids	Spiders
Carabids	Stophylinids

Glossary

Antisense RNA: RNA complementary to the coding or m-RNA strands. This technology provides resistance either by inhibiting gene expression or viral replication.

Antiserum: The blood serum of a warm-blooded animal that contains antibodies.

Avirulence: Pathogenic organism is incapable of causing severe disease.

Acquisition feeding period: The time for which a virus-free vector, actually feeds on a virus infected plants to acquire the virus.

Acquisition access period: The time for which a vector is allowed to feed on a source of virus.

Anastomosis: Fusion between genetically different hypha.

Antibiosis: Liberation of antibiotic or other chemical substances produced by one micro-organism and harmful to the other pathogens.

Antiviral proteins: Novel biomolecules identified from virus resistant non-host plants which provide natural protection to these plants. For *e.g.* Tomato and tobacco plants transferred with Pokeweed antiviral protein (PAP/PAPII) genes exhibits resistance against a number virus and fungal pathogens.

Antagonist: It is a micro-organism that adversely affects another organism.

Anabolism: The process building up body tissue.

Anamorph: The asexual or imperfect stage especially in fungus.

Aggressiveness: Capacity of a pathogen to invade and grow in its host plant and to reproduce on or in it.

Anthracnose: A disease that appears as black sunken leaf, stems, fruit and pods lesions (ulcer like lesions) and caused by fungi that produce their asexual spores in an acervulus.

Acervulus: The mat of hyphae which give rise to closely packed, short conidiophores. It is a characteristic of the Melanconiales.

Antisporalant: A substance (chemical) inhibits spore production without affecting the growth of vegetative hyphae is called antisporalant.

Autoecious: A parasite that can complete its entire life cycle on the same host, term used in rusts.

Aeciospores: The binucleate rust spores produced in an aecium.

Aecium: The cup shaped structure (of binucleate hyphal cells) producing spore chains by conjugate division of the nuclei.

Aerobic: Microorganism living in presence of molecular oxygen.

Agar: A gelatin like substance obtained from sea weed (*Gelidium, Gracillaria, Gigartina*) and used to prepare culture media.

Alternate host: One of the two kinds of plants on which a parasite must develop for completion of its life cycle.

Aplanospore: The non-motile spore.

Apothecium: An open cup or saucer shaped ascocarp of some ascomycetes.

Appressorium: The flat hyphal body or germ tube (swollen tube), enters the epidermal cell of the host.

Ascocarp: The fruiting body of ascomycetes, containing or bearing asci.

Ascogonium: The female gametangium (sex organ) of the ascomycetes.

Ascospore: A sexually spore formed in an ascus.

Ascus: A sac like body (cell) containing ascospores formed as a result of karyogamy and meiosis.

Axenic: A culture of an organism in which no other organism are there.

Auxin: A substance which regulates plant growth and controls cell elongation.

Anisogamy: Copulation between two dissimilar gametes, one smaller (male) and the other bigger (female).

Anaerobic: Relating to a microorganism that lives or a process that occurs in the absence of molecular oxygen.

Asexual reproduction: Type of reproduction which does not involve union of gametes (nuclei) or sex organ or sex cells.

Albinism: Due to lack of light normal color changes into white.

Atrophy: Wasting away of a plant part.

Biotype: A subdivision of physiologic races, sub race or a biotype is a population of individuals which are genetically identical, produced by variant.

Biotrophs (obligate parasite): An organism which always obtained their food in nature from living tissues on which they complete their life cycle.

Bacteriocins: Antibiotic like compounds (bactericidal substances) which are produced by certain strains of bacteria and active against one or more strains of the same or closely related species.

Biological control: Biological control brings about reduction in activity of a pathogen (reduction of the amount of inoculums) by another organism. This may be biocidal (one organism kill the others) or biostatic (the organism only inhibit the pathogen).

Blotch: A disease characterized by large and irregular in shape, spots or blots on leaves, stems and shoots.

Blight: A disease characterized by general and rapid killing (extensive necrosis) of leaves, flowers, stems, branches and twigs.

Bacteriophage: Viruses that attack bacteria are called bacteriophages (or only phages), which means "bacteria eater".

Binary fission: The cell divides in transverse plane into two cells or splitting of cell into two daughter cells.

Biotechnology: The manipulation, genetic modification and multiplication of living organisms through tissue culture and genetic engineering for production of improved or new organism and products *e.g.* crop improvement, industrial improvement and disease control.

Bioinformatics: A mean to store, analyze and interpret the large amounts of data generated. By this combination with information science, genomics may and will help to move biology from *in vivo* to *in silico*.

Bacterium: The unicellular, achlorophyllous, microscopic prokaryotic organism, multiplying by fission.

Basidiocarp: A fruiting body that bears basidia.

Bactericide: A chemical that kills bacteria.

Basidiospores: A sexual spore borne on basidium, following karyogmy and meiosis.

Basidium: A cup shaped structure on which basidiospores are borne.(Typically four).

Bitunicate: An ascus with inner and outer walls.

Budding: The production of a small outgrowth (bud) from a parent cell, a method of asexual reproduction.

Ballistospore: Any spore that is forcibly discharged.

Bioassay: The use of a test organism to measure the relative infectivity of a pathogen or toxicity of a substance.

Biovar: A 'variant prokaryotic strain', that differs physiologically and/or biochemically from other strains in a particular species.

Blister: A raised lesion on the leaf surface with opens to expose spores.

Binucleate: Having two nuclei.

Biciliate: Bearing two cilia.

Biosafety: Biosafety regulations cover assessment of risks and the policies and procedures adopted to ensure environmentally safe applications of biotechnology.

Boomerang effects: Diseases are commonly exacerbated by over-watering and over-fertilization by well meaning gardeners. Efforts by plant pathologists to control disease which result in worse disease as boomerang effects. A classic example of the boomerang effect occurs if a soil-borne pathogen reinvades fumigated soil. Without natural enemies to reduce its spread it often becomes far worse than it would have been before fumigation.

Conducive soil: Soil-borne pathogen develops well and cause severe disease in some soil.

Competition: Competition is a condition in which there is a suppression of one organism as the two species struggle for limiting quantities of nutrients, oxygen, space or other requirements.

Cross protection: The phenomenon in which plant tissues infected with one strain of a virus are protected from infection by other strains of the same virus.

Cellulose: It is a polysaccharides but it consist of chains of glucose molecules and found in plant cell walls.

Cellulase: An enzyme that breaks down cellulose.

Catabolism: The break down phase of metabolism.

Chlorosis: Normal color changed into yellow due to low temperature, lack of iron, excess of lime or alkali in the soil and infection by virus, fungi, and bacteria.

Cell: All living organisms are composed of characteristics type of structural units called cell, cell is the basic unit of life.

Cell wall: The non-living component of a cell (permeable structure). It provides a mechanical support and gives definite shape and protection to the cell, it also helps in the movement of water and solutes towards protoplasm.

Cyanophage: The viruses that attack blue green algae are called cyanophages.

Conjugation: Two compatible bacterial cell come in contact and interchange of genetic material take place of both cells is altered.

Callus: Parenchymatous tissues or masses of cell (thin-walled undifferentiated) developed as a result of wounding or culture on nutrient media.

Cell culture: Growth of any cell.

Clone: The group of genetically identical individuals produced asexually from one individual.

Clamp connection: Outgrowth of fungus hypha which forms bridge connecting hypha cells.

Capsid: The protein coat of a virus particle, consisting of capsomeres.

Capsomere: The protein subunit or small protein molecule of capsid.

Cleistothecium: An entirely closed ascocarp.

Conidium (pl. conidia): A non-motile asexual spore usually formed at the tip or side of a specialized hypha.

Conidiophore: A simple or branched hypha bearing one or more conidia.

Chemotherapy: Control of diseases with chemotherapeutic agents (chemicals).

Chimeric genes: DNA sequence including regulatory and coding sequences not found in nature (together).

Chlamydospore: The thick walled spore formed by modification of hyphal cell, an asexual spore.

Chronic symptoms: Symptoms appearing over a long period of time.

Cistrons: The sequence of nucleotides within a certain area of DNA or RNA.

Cloning: Isolation and multiplication of individual gene sequence by its insertion into a system where it can replicate.

Cloning Vector: A cloning vector is a DNA molecule that has an origin of replication and is capable of replicating in a bacterial cell.

Culture medium: The prepared food material for culturing microorganisms.

Culture: To grow the microorganism artificially on the prepared food material, or the colony of microorganisms on such food material.

Cytokinin: A group of plant growth regulating substances that regulate cell division.

Chromosis: The normal color changed to red or orange are known as chromosis.

Control: Prevention of or reduction of loss from plant diseases.

Canker: A necrotic, often sunken lesion (warty outgrowths) is formed on leaves, stems, branch or twigs and fruits.

Cis-genic plants: made up of using genes, found within the same species or a closely related one, where conventional plant breeding can occur. Some breeders and scientists argue that cis-genic modification is useful for plants that are difficult to cross-breed by conventional means (such as potatoes).

Disease: Any disturbance brought about by a pathogen or an environmental factor which interferes with normal food uptake, synthesis, translocation and metabolism in the plant. In such appearances, yield loss is more than normal healthy plants of the same variety.

Disease trading: When one disease is controlled but another is exacerbated by the same treatment.

Disease cycle: The chain of events involved in the disease development including the stage of development of pathogen and effect of disease on the host.

Disease syndrome: Sum total of all symptoms characterizing a disease or collectively called as syndrome, *e.g.* necrosis of tissue, hypertrophy as well as wilting, total all symptoms called disease syndrome.

Dispersal: The spread of a plant pathogen within the general area or into new geographical area in which it is established is termed dispersal.

Disinfectant: A substance which destroys pathogenic microorganisms.

Disinfestant: A substance which can kill or inactivate pathogenic microorganisms in the environment or on the surface of plant parts before infection.

Diploid: Containing the double (2n) number of chromosomes.

Dissemination: Transfer of inoculums from a diseased plant to a healthy plant or from one place to another.

Disorder: A harmful non-infectious plant disease due to abiotic such as adverse soil conditions, *e.g.* mineral deficiency or toxicity, genetic anomaly, low temperature injury *etc.*

Bdellovibrio: Bdello derives from the Greek word for 'leach'. A genus of gram negative, chaemoorganotrophic bacteria, which parasitizes other bacteria and cause lysis. It multiplies inside the host cell wall and is generally obligate parasites.

Damping off: A disease that rots seedlings at soil level, resulting in the seedling falling over on the ground, or rapid death and collapse of very young seedlings.

Dieback: Progressive death of shoots, branches and roots generally staring at the tip, or extensive necrosis of twigs at their tips.

Disease incidence: The frequency of occurrence of a disease (per cent disease in a crop field).

Disease severity: Per cent disease (amount or severity) in an individual plant.

Disease prevalence: Per cent disease in a defined geographical area.

Decline: Plants growing poorly, leaves small, brittle, yellowish or red, some defoliation and dieback.

Dioecious: Separate male and female organisms.

Dikaryotic: Cell containing a dikaryon.

Disease escape: A phenomenon in which a particular environmental factor prevents infection and disease development in spite of presence of the susceptible host and the pathogen of sufficient pathogenicity.

Elicitors: The cell wall component of the pathogen which are capable of inducing phytoalexin synthesis. Phytoalexins production is stimulated in the most by the presence of certain pathogenic substances called elicitors. Elicitors are high molecular weight substances. It is a constituent of fungal cell wall such as glucans, chitosan, glycoprotein and polysaccharides *etc.*

Epidemiology: The study of epidemic and factors that influencing them is called epidemiology. Three infection chains are part of epidemiology, (i) survival (ii) dispersal and (iii) infection.

Epidemic or epiphytotic disease: An epidemic or epiphytotic disease usually occurs widely but periodically in a destructive form. The pathogen may be present as in endemic diseases but the environmental factors responsible for the development of the disease occurs only periodically, or A wide spread and severe outbreak of disease. It occurs widely but periodically in a destructive form.

Endemic disease: When it is constantly present in a moderate to severe form in a particular country or district, *e.g.* wart disease of potato (*Synchytrium endobioticum*) is endemic in Darjeeling.

ELISA (enzyme linked immune sorbent assay): It is a serological test in which the antibody carrying the enzyme, that releases a colored compound.

Endoparasite: A parasite which inters the host and feeds from within the host.

Eucarpic: Reproductive structure formed certain parts on the thallus.

Endobiotic: An organism living within its substratum.

Ectoparasite: A parasite which feeds on host from the exterior.

Epibiotic: An organism whose reproductive organs are on the surface of the substratum.

Exanthema: Eruption or discharge of gum or other substances from diseased tissues.

Epidemic rate: The amount of increase of disease per unit of time in a plant population.

Epiphytically: Existence on the surface of a plant/plant organ without causing infection.

Etiology: Study of the cause of the plant disease and nature of the causal agent.

Eradicant: A chemicals that destroys a pathogen at its source.

Eradication: Control of plant diseases by eliminating the pathogen after it is established.

Exudate: A natural liquid discharge from plant organs or a liquid discharge from diseased tissue.

Exclusion: Preventing new pathogens and diseases from reaching an uninfected area and avoiding contact between the pathogen and the crop or field, or control of plant diseases by excluding the pathogen or infected plant material from disease free areas.

Ethylene: It is gaseous hormone increase during invasion of pathogen in the plants.

Etiolating: Yellowing of plants due to lack of light.

Enzyme: Enzymes are large protein molecules that are catalyzing all interrelated reaction in the living cell.

Ergot: It is a fruiting body by a fungus, *Claviceps purpurea* which is parasitic on rye and other grasses.

Embryo: The rudimentary seedling.

Endosperm: The food storage for nourishing the embargo.

Externally seed borne: Some pathogen carried on the seed, *e.g.* covered smut of barley (*Ustilago hordei*), brown spot disease of rice (*Drechslera oryzae*), covered smut of oat (*Ustilago kolleri*), black scurf of potato (*Rhizoctonia bataticola*) and grain smut of jowar (*Sphacelotheca sorghi*).

Explant: Plant parts called explant, used as initial inoculum *e.g.* protoplast, tissue, organ *etc.*

Enation: Tissue malformation or overgrowth induced by certain virus infections.

Escape: Certain crop, varieties which undergo development and maturation may complete their life-cycle before maximum infection.

Episome: In bacteria, a DNA molecule, other than the bacterial chromosome, capable of replication.

Epinasty: Downward bending of a petiole, lamina and stems or lamina may continue the curve of the petiole so that its upper surface faces inwards towards the stems.

Facultative parasites: Some organism lived most of the time on dead organic matter, but under certain condition may attacks living plants and become parasite.

Facultative saprophyte: Some organism lived most of the time as parasite but under certain condition may grow saprophytically on dead organic matter.

Fungistasis: Phenomenon wherein viable fungal propagule do not germinate in the soil in condition of temperature and moisture favorable for germination.

Fumigation: Many chemicals which can produce toxic gases are used to kill the pathogen.

Forma specialis (f. sp.): A group of biotypes of a pathogen species.

Fumigants: Toxicants which are used in gaseous form for killing insect pest and microorganism and act as soil sterilant, *e.g.* methyl bromide, methane, CO_2, chloropicrin, tetrachloroethane *etc.*

Fungi: Fungi are eukaryotic organism, spore bearing, chlorophyllus, filamentous and branched, they reproduce by sexually and asexually means.

Fungicide: A substance (chemical) capable of killing fungi.

Fungicidal: Killing fungal spores or mycelium by physical agents such as heat, ultraviolet light, x-rays, gamma radiation as well as chemicals.

Fungistat: A substance (chemical) is preventing the growth of a fungus without killing it.

Filtration: Filtration is the sterilization process, by which removal of organism of particulates from thermo labile solutions, air, waters.

Freeze drying: A technique used in the preservation of plant tissues and microorganism *etc.*, whereby water is removed under vacuum while tissue remains in the frozen state. Synonymous with lyophilization.

Filiform: Thread like.

Fusiform: Spindle shaped, narrowing toward the ends.

Fission: Asexual reproduction of bacteria involving transverse splitting of the cell or the splitting of a cell into two cells.

Flagellum: The whip like locomotion organ originating from bacterium or zoospore.

Fructification: Production of spores by fungi.

Fruiting body: A complex fungal structure containing spores.

Fasciation: A plant disease, due to cell injury in the bud, resulting in flattened and sometimes spirally curved shoots.

Fermentation: Oxidation of certain organic substances in the absence of molecular oxygen.

Fertilization: The sexual union of two nuclei or protoplasts resulting in doubling of chromosome number (2n).

Gummies: Emission of gum by or in a plant tissue or organ.

Gum: A complex polysaccharide formed by cells in reaction to wounding or infection.

Growth regulator: A hormone (chemical) which synthesized in the particular cells and initiates the growth and reproduction of the plant.

Growth retardant: A substance that decreases the rate of growth of a fungus, higher plant *etc.*

Growth inhibitor: A substance that inhibiting the growth of an organism.

Gene-for-gene hypothesis: Every resistance and susceptible gene present in the host while pathogen has a gene for virulence and virulence. Susceptible reaction would result only when the pathogen is able to match all the resistance genes present in the host with appropriate virulence gene. If one or more resistance genes are not matched by the pathogen with the appropriate virulence genes, resistance reaction is the result. This complimentary relationship of the host and pathogen is called gene for gene hypothesis for disease resistance.

Germination: The start of growth of the spore or seed or the emergence and development of seedling form the seed embryo which able to produce normal plant under favorable condition.

Genome: A complete set of chromosomes of a diploid species or single set of chromosome, or all genes together called genomes.

Genetic engineering: The alteration or manipulation of genetic composition of a cell by various procedures such as transformation, protoplast fusion *etc.* in tissue culture.

Gene: Linear portions of the chromosome. Fictionally, gene is unit of inheritance and structurally, gene on a segment of DNA which codes for one polypeptide.

Gram-positive: Bacterial cell retaining the Gram's stain after washing with alcohol, appear dark blue or violet.

Gram-negative: Bacterial cell loss the stain after treatment of alcohol, loss of crystal violet color.

Germplasm: The sum total of all hereditary material or genes present in a species.

Gibberellin: A group of plant growth regulating substances.

Gamete: A differentiated sex cell or a sex nucleus that fuse with another in sexual reproduction or the male or female sex cell or a sex nucleus.

Gene cloning: Isolation and multiplication of an individual gene sequence by its insertion into a bacterium.

Genus: A taxonomic group including species.

Guttation: Exudation of water from plants, particularly along the loaf margins.

Gall: Outgrowth or swelling, often more or less spherical produced on a plant as result of attack by bacteria, fungi or other organisms.

Hypersensitivity: Excessive sensitive of plant tissues to certain pathogens. Affected cells are killed quickly, blocking the advance of obligate parasites, or the extreme degree of susceptibility in which rapid death of the cells in the vicinity of the invading pathogen (the infection court) occurs. This halts/ stop the progress of the pathogen although it may not die immediately. Thus, hypersensitivity is a sign of very high resistance approaching immunity, *e.g.* rust, virus and nematodes.

Host: A living organism harboring a parasite.

Hemibiotroph: A parasites or an organism which may attack living tissues in the same way as biotrophs but continue to develop and speculate after the tissue is dead, *e.g.* leaf spotting fungi.

Horizontal resistance (uniform or non-specific resistance): Partial resistance equally effective against all races of a pathogen.

Hypovirulence: Reduced virulence of a pathogenic strain as a result of the presence of transmissible double stranded RNA. The hypovirulence strain carry virus like dsRNA. The double stranded RNA apparently passes through mycelia anastomoses from hypovirulent to virulent strains and reduced the development of canker. Chestnut blight (*Cryphonectria parasitica*) control with hypovirulent strains of the pathogen.

Hyperparasite: A parasite parasitic on another parasite.

Hemicellulose: A complex mixture of polysaccharide resembling cellulose and main component of cell wall. It is insoluble in water, which are attached with cellulose and lignin.

Hypertrophy: Excessive growth (abnormal increase) in the size of individual cell (of affected organ).

Hyperplasia: A condition in which a tissue is enlarged as result of excessive production of cell (increase in the number of cell).

Helper virus: A virus capable of supporting the replication of an additional, nonessential and unrelated nucleic acid component, *e.g.* a virusoid or satellite RNA.

Hybridization: In fungi, the crossing of two individual differing in one or more heritable characteristics, or combination of dissimilar gametes.

Heterokaryosis: A condition in which genetically different nuclei are associated in the same protoplast or same mycelium.

Hypha: A single branch of mycelium, or the unit o structure of most fungi, tubular, filaments.

Hfr: In *Escherichia coli*, high frequency recombination; cells in which the F factor is integrated into the bacterial chromosome.

Hyperauxiny: A condition in plant in which the level of auxin (IAA) in the tissues is higher than normal. Hyperauxiny may result from infection by fungi, bacteria, viruses and nematodes and by chemical and environmental stimuli.

Heterogamy: The fusion between male motile gametes and non-motile female gametes.

Haustorium: The penetrating and absorbing organ of the hypha of a parasite.

Heteroecious: A parasitic fungus requiring two host species for the completion of its life cycle.

Holocarpic: Whole thallus converted into zoosporangium or reproductive structure.

Homothallic: Fungus with compatible male and female gametes produced on the same mycelium.

Hyaline: Without any color.

Hermaphroditic: The individual having both functional male and female sex organ.

Hydathodes: Structures with one or more openings that discharge water from the interior of the leaf to its surface.

Heterokont: A biflagellate structure with two flagella, unequal in size.

Haploid: Containing reduced (n) number of chromosomes.

Habitat: The natural place of occurrence of an organism.

Hypoplasia: Under development of tissue due to decrease in cell division.

Herbicide Resistant Plants: Plants that can tolerate herbicides.

Induced resistance: Enhancement of the plants defensive capacity against a broad spectrum of pathogens and pests that is acquired after appropriate stimulation.

Inoculum: Any part of the pathogen which cause disease, *e.g.* spore, mycelium, chlamydospore, sclerotia *etc.*

Inoculation: The transfer or arrival of a pathogen on to a host.

Inoculate: To bring a pathogen into contact with a host plant or plant organ.

Infection: The establishment of a parasite within a host plant.

Invasion: When the pathogen rapidly grows in the host tissues called invasion.

Inoculum density: It is a mode up of fungal propagules having a certain amount of virulence and vigour.

Infection concert: Place where infection may take place.

Isogamy: Copulation between two gametes morphologically same.

Infections disease: A pathogenic disease spreading from diseased to healthy plant.

Isogametes: The two gametes (male and female) are similar in size.

Isogametangia: Gametangia, presumably of opposite sex, that are morphologically indistinguishable.

Inoculum potential: Combined energy of fungal propagules (or other microorganism) to cause infection.

Incubation period: The period of time between penetration of a host by a pathogen and the first appearance of symptoms on the host.

Inoculation access period: The time for which the vector, after acquiring the virus is allowed to feed on healthy plants and transmit the virus.

Inoculation feeding period: The actual period of feeding is called inoculation feeding period.

Inoculation threshold period: The minimum time necessary for a vector to feed on test plant in order that transmission may occur.

Intellectual Property Right (IPR): Legal rights of a person, grp or institution to prevent or protect intellectual property (innovations, discovery, concept, knowledge *etc.*).

Immune: Exempt from disease, cannot be infected by a given pathogen.

Immunity: Exempt (100 per cent) freedom from infection. It denotes that the pathogen cannot establish parasitic relationship with the host, even under the most favorable conditions, or the state of being immune.

Isolate: A single spore or pure culture and the subcultures derived from it and also used to indicate collection of a pathogen made from different places or at different times.

Isolation: The separation of a pathogen from its host and its culture on a nutrient medium.

Integrated control: To use all available methods of control of disease/pests.

Integrated disease management (IDM): The simultaneous, use of a number of strategies in the field to minimize the loss caused by plant diseases. Field sanitation, host tolerance, agronomic and cultural practices and the use of biological agents can be pressed into service to keep the loss caused by plant diseases at low level.

Injury: Damage of a plant by animal, physical or chemical agent.

Internally seed borne: Some pathogen carried inside the seed, *e.g.* loose smut of wheat (*Ustilago nuda tririci*), loose smut of barley (*U. nuda*), stripe disease of barley (*Drechslera graminea*) and red rot of sugarcane (*Colletotrichum falcatum*).

Imperfect fungus: A fungus that is not known to produce sexual spores or one lacking any sexual reproductive state.

Imperfect stage: No sexual spores are produced; anamorphic.

Intercellular: Between cells.

Intracellular: Within or though the cells.

In vitro: When biological processes are made to occur outside the organism in a vessel or test tube or in culture, outside the host.

In vivo: Within the living organism, or in the host.

Intercalary: Between apex and base.

Karyogamy: The fusion of two nuclei resulting in the production of diploid zygote nucleus.

Koch's postulates: Three criteria proposed by R. Koch's for proving pathogenicity.

Life cycle: The successive stages in the growth and development of an organism that occur between the appearance and reappearance of one stage of the organism (*e.g.* spore, eggs *etc.*).

Latent period: The time between inoculation and the start of sporulation by the resulting infection, or the period required by a virus to multiply and accumulate within the vector before the latter become infective.

Latent infection: Host is infected but does not show symptoms.

Latent virus: A virus that does not induce symptom development in its host.

Lipid: Substances (organic compound) whose molecules consist of glycerin and fatty acids and sometime certain additional types of compounds.

Lignin: A complex organic substances or group of substances that is found in the middle lamella in the cell wall and in the cell wall of xylem vessels.

Leaf spots: Necrosis of the tissues around the substomatal space and appear brown or water soaked and spots usually remain restricted in the growth.

Line: A group of individuals having common parents or ancestors.

Lyophilization: A technique used in the preservation of microorganisms, plant tissues *etc.* at -70°C and then dehydration by vacuum.

Leaf curl: Distortion, thickening and curling of leaves *e.g.* along the two sides of midrib.

Leaf roll: Curving of the leaf lamina, generally towards and parallel with the midrib.

Lysogeny: The chromosome of a bacteriophage becoming integrated into the chromosome of its host, in case of temperate phages, such bacteria are called lysogeny.

Lesion: A localized area of the diseased or discolored host tissue.

Local lesion: A host, which develops local lesions on inoculation with virus.

Lichen: A combination of an alga and a fungus in which the two components are so interwoven as to form what appears to be a single individual, benefiting each other.

Lysis: A breaking down or dissolution of cells by enzymes or bacteriophage.

Monoculture: Continuous cultivation of the same crop in a conducive soil. After some year, severe disease eventually leads to reduction in disease through increased population of microorganism antagonistic to the pathogen.

Mycorrhiza: The mycelial hyphae living in association with roots of higher plants. The nature of this association is believed to be symbiotic (mutualism) non-pathogenic or weekly pathogenic.

Morphovars: Strains that differ physiologically.

Monogenic: A character governed by single gene with 3:1 Mendelian ratio *e.g.* resistance of oats is governed by single gone.

Meiosis: A series of two nuclear division usually in quick succession in which chromosome number is reduced by one half.

Maceration: The separation of cells along the line of the middle lamella is called maceration. The term 'cell separation' is more appropriate, or softening of the tissues or loss of plant tissue by pectic substances.

Metabolism: Chemical reaction occurring in the living cell by which the food is transformed into living protoplasm; energy for the activities of the body.

Mosaic: Normal or light green yellow patches symptoms on the plant surfaces produced by certain virus diseases.

Mottle: An irregular pattern of distinct light and dark areas.

Messenger RNA: A chain of ribonucleotides that codes for a specific protein.

Microscopic: Very small can be seen only under microscope.

Mildew: A fungal disease of plants in which the mycelium and spores of the fungus are seen as a whitish growth on the host surface, or area of stem, leaves, fruits covered with whitish mycelium.

Mold: Any profuse or wooly fungus growth on damp or decaying matter or on surfaces of plant tissue.

Malformation: A condition in which growing shoots become stunted and dwarf with leaf buds flowers suppressed and crowded presenting a bunchy top appearance.

Mushroom: A fleshy, sometimes tough umbrella like sporophores that bears basidia on the surface of gills.

Micro-organism: The unicellular or multicellular microscopic organism and widely distributed in air, soil, water, the living plant bodies and animals and dead organic matter, *e.g.* fungi, bacteria, viruses, yeast, mold, algae, protozoa *etc.*

Monocyclic: Having one cycle per season.

Mutation: Sudden change of genetic material of a cell.

Mycoplasma like organism (MLO): A prokaryotic organism, lack of cell walls, are bounded by unit membrane, have cytoplasm, ribosomes and strands of nuclear material, pleomorphic shape and present in the sap of a small number of phloem sieve tubes and assumed to be the causes of the disease, they resemble mycoplasma in all respects except that they cannot yet be grown on artificial nutrient media.

Mycoplasmas (true mycoplasmas): Pleomorphic, prokaryotic organism, lack of true cell wall, bounded by unit membrane (triple layered), reproduce by binary fission and budding of cells, have no flagella, produce no spores and are gram-negative and all true mycoplasma parasitic to humans and animals and all saprophytic ones can be grown on more or less complex nutrient media.

Mycelium: Mass of hyphae that make up the body of a fungus.

Mycosis: An infection by a parasitic fungus, or a diseased so caused.

Mycotoxin: Toxic substances produced by fungi which forms toxins.

Monoecious: Male and female reproductive organs in same individuals.

Monokaryotic: Having single nucleus.

Molecular markers: Nucleic acid segments that behave as landmarks for genome analysis and are based on naturally occurring genetic variability (usually termed polymorphisms).

Marker assisted selection (MAS): Indirect selection for a primary trait (trait for economic importance) via direct selection for a secondary trait (marker), whch is independent of growth stage, environment, and gene - environment (G x E) interaction.

Mycotoxicosis: Disease of animal and human caused by consumption of feed and foods invaded by fungi that produce mycotoxins.

Metachromasy: The phenomena in which certain biological entities (*e.g.* metachromatic granules) become stained with a color which is different to that of the dye used to stain them.

Medium: Substratum employed in the laboratory for growing microorganisms.

Microcyclic: Rusts having teliospores as the only binucleate spores (also called short cycled) in the life cycle.

Micropropagation (clone propagation) - It is a process of production of clones similar to asexual reproduction. Plantlets are produced from shoot tips/ axillary buds on culture medium omitting callus phase. Here, small amount of explant produce millions of clonal parts in a year directly.

Monocentric: Thallus bearing single reproductive structure (sporangium or resting spore).

Monoclonal antibody: Continuous cultures of fused cells secreting antibody of predefined specificity.

Mycology: Study of fungus (fungi) is known as mycology.

Molecular plant pathology: Field of biology and plant pathology that studies the structure and function of genes at a molecular level. It studies, what genes are involved in plant pathogenesis and defenses against diseases, employing the methods of plant pathology and molecular biology.

Native/wild type genes: Gene that is present in the genome of an untransformed cell *i.e.* a cell not having known mutation.

Necrotroph (perthotrophs or perthophytes): These are organisms which kill the host tissues in advance of penetration and then live saprophytically.

Necrosis: Death of the cells, tissues, organs or entire plants (loss of tissues).

Necrotic: Death and discolored of the tissues and other plant parts.

Nucleotide: Made up of three essential components: a nitrogenous base, a sugar, and a phosphate moiety.

Non-persistent virus (stylet borne or proboscis virus): Non-persistent virus survived in their vectors for only short period, less than the survival time of virus in untreated leaf extract.

Natural openings: Stomata, lenticels and hydathodes are natural openings in plants.

Net blotch: Dead leaf spot with a net-like pattern of dark veins.

Non-infectious disease: A disease that is caused by an abiotic agent, not by a pathogen.

Non-septate: Without coos walls.

Nucleocapsid: Nucleic acid covered with protein coat (capsid).

Nucleic acid: An acidic substance which contains phosphorous, pentose, pyrimidine and purine base, it determine the genetic properties of an organism and viruses.

Nucleoprotein: Consisting of nucleic acid and protein.

Oligogenic: A character governed by several genes, *e.g.* dihybrid, trihybrid; anthracnose of bean is governed by three gene.

Obligate parasite: An organism that can obtain food only from living organism; cannot be grown in culture media.

Obligate saprobe: An organism that must obtain its food from dead organic matter and is incapable of infecting another living organism.

Oogonium: A female gametangium of oomycetes containing one or more gametes (eggs).

Oospore: A thick-walled sexual spore produced by the two morphologically different gametangia (oogonium and antheridium).

Oosphere: A naked female gamete which is non-motile.

Ovary: Female reproductive body producing or containing eggs.

Ozone (O_3): A highly reactive form of oxygen which in relatively high concentration may injure plants.

Oidium: A thin-walled, flat-ended asexual spore formed by breakup of hypha. It behaves as a spore or spermatium.

Oogamous: It refers to a type of fertilization in which two heterogametangia come in contact with each other and contents of one flow into the other through a pore or tube.

ORF (Open reading frames): Amino acid sequences encoded between translation initiation and termination codons of a coding sequence.

Organogenesis: Process of differentiation of callus initially into embryo like structure (embryoids) which further develop into organs like roots/shoots.

Ostiole: It is neck-like structure in an ascocarp; lined with periphyses, and terminating in a pore; also the opening of a pycnidium or perithecium which is pore-like.

Pathogenicity: The ability (capability) of pathogen to cause disease.

Pathogen: An entity (parasite or causal agent) able to cause disease in a particular host or range of host.

Pathogenesis: The chain of events that leads to disease development in the host. It involves the action of pathogen, susceptibility of host and affects of environment.

Parasite: An organism which derived the food materials they need for growth from living plants (the host or suspect) are called parasite. Most but not all, pathogens are parasites and similarly most but not all parasites are pathogens.

Parasitism: The relationship between host and parasite is known as parasitism.

Penetration: Entry of the pathogen (an entity) into the host tissues or the initial invasion of a host by a pathogen.

Photosynthesis: A process, by which chlorophyll containing organisms synthesized carbohydrates from CO_2 and water using the energy of sunlight.

Plasmid: A self-replicating extra chromosomal genetic structure (circular dsDNA molecule) found in certain bacteria and fungi; generally not required for survival of the organism.

Physiologic specialization: The occurrence of physiologic races within a species.

Physiologic race: Pathogen of the same species and variety which cannot be distinguished by their structure, but differ in physiological behavior, particularly in their ability to parasitize varieties of a given host.

Protoplast fusion: Fusion of protoplast from two different plant species and regeneration of somatic hybrid plants (disease resistant plant). Protoplast fusion also called somatic hybridization.

Pathovar: A subspecies or group of strains in bacteria that can infect only plants within a certain genus or species.

Pathotype: A subdivision of a species distinguished by the common characters of pathogenicity, particularly in relation to the range of host.

Physiotype: A population of pathogen in which all individuals have a particular character of physiotype (but not pathogenicity in common).

Pathogenic: Capable of causing disease in a host or range of hosts.

Persistent or circulative virus (propagative virus): Persistent virus survived in their vectors for long periods, sometimes a weeks or months.

Parasexualism: A process in which plasmogamy, karyogamy and haplodization take place in sequence, but not at specified points in the life cycle of an individual.

Phytoalexins: Low molecular weight, antimicrobial compounds, formed by plants in response to infection and exposure to various toxic and non-toxic compounds and is thought to play a role in disease resistance.

Plant quarantine: Preventing entry of pathogen from infested area into new non-infested area.

Plant protection: A technique of safeguarding growing crops and storage of agricultural products from the natural enemies.

Prototroph: A microorganism which is able to grow normally or on minimal medium containing inorganic salts and a carbon source.

Phytopathogenic: Micro-organism inciting plant disease.

Perfect stage: The sexual stage (fruiting bodies) in the life cycle of a fungus. The teliomorph.

Propagule: A part of an organism that may be disseminated and reproduce the organism; or one unit of inoculum of any pathogen is called propagule.

Pathodeme: Pathodeme is that population of a host in which all individuals have a particular character of resistance in common.

Predisposition: The effect of one or more environmental factors which makes a plant vulnerable to attack by pathogen. It is process which antedates penetration and infection.

Pasteurization: A heating process that destroys pathogenic bacteria in a fluid such as milk, cheese, vine, beer, fruit juice *etc.* below 100°C, mostly at 62.8°C for 30 minutes.

PPM (parts per million): An amount of substance/compound contain in a million part of mixture or solution in carrier such as air, water. One part per million equals one milligram in one kilogram.

Polycentric: Thallus bearing more than one reproductive structure (sporangia or resting spores).

Paraphyses: Sterile, basally attached structures in a hymenium.

Periphyses: Short hair-like growths in ostiole or pore in the stroma.

Primary inoculum: The inoculum that survives the winter or summer and cause the original or emotional infection in the spring or autumn is called primary inoculum.

Primary infection: The first infection by a pathogen after it has gone through a period of resting or dormancy; or the first infection of a plant by overwintering or over summering pathogen.

Pandemic disease: These occur all over the world and result in mass mortality.

Phyllosphere: Surrounding area of the leaf surface or plant surface, where the micro-organisms two types of interaction take place, synergistic and antagonistic.

Protectant: A substance (chemical) which is effective only if applied prior to fungal infection, or a substance that protects an organism against infection by a pathogen.

Perithecium: A closed ascocarp with an opening or pore (ostiole).

Phenotype: The externally visible appearance.

Planogametic copulation: Fusion of naked gametes, either one or both is motile.

Precipitin: An antibody capable of causing precipitation of soluble antigen.

Promycelium: The germ tube from teliospore bearing the basidiospores; the basidium.

Protoplast: A naked cell or cell without cell wall.

Polygenic: A character controlled or governed by many genes. It involved in a large number of diseases such as cotton wilt.

Plant Pathology or phytopathology: It consists of three Greek words (phyton-plant, pathos- ailments or diseases and logous-discourse or knowledge). Plant Pathology is the science which deals with causes, symptoms, etiology, perpetuation, losses of plant diseases and their management.

Pest: Pests are organism that cause economic losses.

Protein: Highly complex chemical substances in pure form containing C, H, O and occasionally phosphorus and sulphur both.

Plant: Living organism with neither the power of movement nor special organs for digestion.

Promoter: A nucleotide sequence, usually upstream (5′) to its coding sequence, which controls the expressions of coding sequence by providing recognition for RNA polymerase. *e.g.* TATA box and other seuqnces.

Protoplasm: A cell is an organized mass of living substances.

Pathotoxin: A toxin produced by a pathogenic organism and capable of simulating the disease.

Phytotoxin: A toxin toxic to plant.

Protoplast: Animal, plant or fungal cell from which the entire cell wall has been removed.

Purification: The isolation and concentration of virus particles in a pure form which is free from cell components.

Pustule: The blister like swelling of epidermis during spore emergence.

Pycniospore: The spore formed in a pycnium; also known as spermatium.

Pycnium: Fruiting body of the rusts producing pycniospores or spermatia; also known as spermogonium.

Polymerase: An enzyme that joins single small molecules into chains of such molecules (*e.g.* DNA, RNA).

Race: A genetically and often geographically district mating group within a species; also a group of pathogens that infect a given set of plant varieties, or physiological race.

Resistance: The inherent ability of a plant to prevent or restrict the establishment and subsequent activities of a potential pathogen; or the ability of an organism to exclude or overcome, completely or other damaging factor; plural- resistance is the process.

Resistant (singular-resistant is the result): Possessing qualities that hinder the development of a given pathogen infected little or not at all.

Rhizosphere: The soil surrounding the roots or the region of soil which is under influence the plant roots.

Rhizoplane: The root surface and its adhering soil are sometimes termed rhizoplane.

Roguing: Removal and destruction of infected plant parts from the field or crop.

Respiration: The oxidation (aerobic, anaerobic breakdown) of organic materials, primary carbohydrates change into H_2O and CO_2 and produce utilizable energy.

Relative humidity: The amount of water present in the air at a given temperature in proportion to the maximum water holding capacity at temperature in atmosphere.

Ribosome: A subcellular article involved in protein synthesis.

Ribozyme: The term 'ribozyme' is referred to the ability of a RNA molecule to cleave target RNA molecules. Although ribozyme specific for any unwanted gene can be designed and integrated into plant genome, this approach has limited use in the development of virus resistant transgenic plants.

Rickettsia like organism(fastidious vascular bacteria): RLO are prokaryotic organism, rod-shaped, they bounded by cell membrane and a cell wall, they have no flagella; nearly all fastidious vascular bacteria are gram-negative only the xylem inhibiting bacteria causing sugarcane stunting and bermuda grass stunting are gram-positive; multiply only in a eukaryotic cells.

Root rot: Disintegration or decay of parts or all of the root system.

Rust: Many small lesions or pustules of spores on leaves, stems, usually of a powdery mass of brownish, red, yellow or dark brown color or rusty color.

Rots: The infected tissue dies, decomposes rapidly and turns brown. Fungi and bacteria which are able to dissolve cell walls cause these symptoms.

Receptive hypha: The special hypha acting as a female gamete or gametangium.

Regeneration: Production of entire plant from explant is called *regeneration.*

Resting spore: The spore resistant to extremes of temperature and moisture, germinating only after a period of time.

Resting stage: The inactive stage.

Ring spot: Symptom of certain virus diseases characterized by circular chlorosis with a green center.

Russet: The rough areas on skin of fruit due to cork formation.

Rosette: Short bunchy growth of plant parts.

Reproduction: The production of new individuals having all the characteristics typical of the species.

Rhizomorph: The thick organized strand of hyphae. The growing tip of strand resembling the root tip.

Recombinant inbred lines (*RILs*): Lines derived by single seed descent from individual F_2 progeny.

Serovars: Strains that have antigenic properties that differ from other strains.

Strain: The descendants of a single isolation in pure culture, an isolate or a group of similar isolates; a race; a group of virus isolates (in plant viruses) having most of its antigens in common with that of another group (strain).

Susceptible: Lacking the inherent ability to resist disease or attack by a given pathogen; non-immune, *i.e.* subject to infection.

Suspect: Any plant or organism that can be attacked by a given pathogen or given disease; a host plant.

Susceptibility: The inability of a plant to resist the effect of a pathogen or any other damaging factor.

Symptom: The external and internal reactions or alternation of a plant as a result of disease. or a visible abnormality arising from disease; signs of infection.

Streaks: Elongated narrow lesions on leaves which are usually of brown shade. *e.g.* barley stripe disease.

Sexduction: In bacteria, the incorporation of bacterial genes into a sex factor (*e.g.* F) and their subsequent transfer to other bacterial cells through conjugation.

Spreader: Spreader is a material which improve the contact between fungicide and sprayed surface; *e.g.* mineral oils, glyceride oil and soap; it reduces the surface tension.

Sticker: A substance added to spray or dust which improve the adherence to plant surface, *e.g.* gum Arabic, dextrins, fish oil, milk casein and gelatin.

Safner: A chemical which reduce the phytotoxicity of another chemicals, *e.g.* the phytotoxicity of CuSO4 is reduced by lime.

Symbiosis: A mutually beneficial association of two or more different kinds of both partners appears to benefit.

Syndrome: The totality of effects produced in a plant by a disease.

Saprophytes: An organism which feeds on dead organic matter; or an organism which derived their nutrition from dead organic matter.

Serotype: A population of a pathogen (usually a bacterium or virus) in which all individuals possess a given character of serology in common.

Serology: A method using the specificity of the antigen-antibody reaction for the detection and identification of antigenic substances and the organisms that carry them.

Secondary inoculum: Inoculum produced by infections (from primary infection) that took place during same growing season. It is responsible for spread of disease because of the primary infection spores and other structure produces secondary infection. The secondary inoculum is spread by insect vectors or by other means.

Secondary infection: Any infection caused by inoculum (secondary inoculum) produced as a result of a primary infection or subsequent infection ex. late blight of potato.

Sporadic disease: These belong to the endemic group. They occur at very irregular intervals and locations and in relatively few instances ex. Angular leaf spot of cotton and blotch disease of cucumber.

Suppressive soil: Soil in which certain disease are suppressed, because of presence in the soil of microorganisms antagonistic to the pathogen.

Survival: In the absence of their cultivated host, pathogen survives some alternate host, the infection will remain incomplete.

Sclerotium: A hard resting body resistant to unfavorable conditions and can germinate under favorable conditions.

Scorch: The burning of leaf margins.

Satellite RNA: An encapsulated RNA component which is dependent upon a helper virus for replication but show no significant sequence homology with either the viral or host genomes, or one which multiples only in the presence of a specific second virus.

Siderophore: Extracellular low molecule weight compound produced by microorganism it has bindable capacity, it binds the iron (Fe^{++}) and make less available to pathogen.

Soil polarization: When solar heating increased temperature in the soil to level above the optimum, viability and activity of pests is relatively affected.

Sanitation: Removal and burning of infected plant parts, decontamination of tools, equipment's, worker's hands *etc.*

Satellite virus: Polyhedral ssRNA viruses which are unable to replicate without the presence of a second or helper virus are called satellite or defective encapsulated in a complete virion. Some plant viruses such as TNV have satellite virus associated with it.

Starch: A glucose polymer or main reserve polysaccharides found in plant cells. It is synthesized in the chloroplast and in the amyloplast or a polysaccharide consisting of glucose units.

Shot hole: Falling off of diseased leaf tissues leaving a hole behind.

Sign: The pathogen or its parts or products visible on the host plant.

Smut: The disease in which infected plants show masses of dark powdery spores caused by the smut fungi (Ustilaginales).

Sporangiophore: A hypha that bears sporangia.

Sporangiospore: A spore borne within a sporangium.

Spore: A minute propagative unit as of fungi which may germinate and develop into a new individual.

Somatic: The vegetative phase, structure or function.

Stress: Predisposition effect of external environment on plants and diseases.

Soft rots: Disintegration of cells and dissolution of middle lamella as a result of enzyme action, very often a dirty liquid oozes out of the affected parts and changes the color or maceration and disintegration of fruits, roots, bulbs, tubers and fleshy leaves.

Seed longevity: Seeds remain viable for many years.

Seed: A mature ovule consisting of a embryo and endosperm and the protective covering (seed coat).

Seed pathology: The science and technology dealing with seed borne plant diseases, the mechanism of their transmission, the method for preventing and controlling of these diseases in the field and during seed storage, the assessment of seed borne inoculum and techniques for disease detection in seeds.

Seedling: A young plant developed from seed.

Somatogamy: Union of somatic cells during plasmogamy.

Sorus: The compact mass of fruiting bodies or spores (mass of sporangia).

Stolon: Hypha growing horizontally along the surface of substrate.

Sterile fungi: Fungi not producing spores.

Sterilization: Complete destruction of all living organisms by physical or chemical agents.

Stroma: The compact mycelial structure, like a mattress, on or in which fructifications are usually formed.

Soil inhabitants: Microorganisms which are able to survive in the soil indefinitely as saprophytes.

Soil invader: Plant parasitic fungi and bacteria that pass a part of their life cycle in soil.

Seed lot: Seed lot is a specified quantity of a seed which is physically identifiable.

Seed sample: A small quantity of seed as representative of a seed lot.

Seed protectant: A chemical applied to seed before planting to protect seeds and new seedlings from disease and insect.

Seed dressing: The process of cleaning seed.

Shoot tip culture: Culture of shoot tip (shoot apical meristem) along with one or more leaf primordia or mature leaves is called shoot tip culture.

Somaclone: Plants derived from somatic cell culture.

Somoclonal variation: Somaclonal variations are heritable variations for both qualitative and quantitative traits produced in plants regenerated from cell culture and tissue cultures.

Spiroplasma: Pleomorphic prokaryotic wall less, microorganisms, bounded by unit membrane, cells varying in shape from spherical or slightly ovoid, helical and branched non-helical filaments (in culture), and inhabiting plant phloem, gram positive in nature and cultured on nutrient media, there are no flagella, reproduce by binary fission.

Synergism: Joint action of two microorganisms resulting in production of an effect not produced by either organism separately.

Systemic: Spreading of pathogen or chemical throughout the plant body internally.

Synnema: A group of conidiophores cemented together and forming an elongated spore bearing structure.

Swarm cell: A flagellated cell; usually applied to the motile cells of the Myxomycetes and the Plasmadiophoromycetes.

Sporodochium: The cushion shaped fruiting body having a bunch or cluster of conidiophores woven together on a mass of hyphae.

Sporangium (pl. sporangia): A sac-like structure, the entire protoplasmic contents converted into an indefinite number of spores.

Spawn: Packed pure culture of a specific strain of edible mushroom, growing on a medium usually grain.

Sclerotia: Hard resting body resistant to unfavorable conditions.

Staining: The process which is used to study the properties of various microorganisms and their differentiation into specific group/genera/species.

Smut: Seed or a gall filled with mycelium, or a disease caused by smut fungi and characterize by appearance of sooty or charcoal like powdery mass on the affected organs.

Scab: Localized lesions or ulcer like lesions (roughened crust like lesion) on the fruits, leaves, tubers.

Tolerance: The ability of a plant to sustain the effects of a disease, without dying or suffering serious injury or crop loss.

Tolerant: Able to endure infection by a particular pathogen without showing severe disease or giving a little reaction to the effect of other factors (*e.g.* a virus tolerant of heat).

Transmission threshold period: The minimum period of time that virus needs for acquisition and subsequent transfer to a virus free plant, is the transmission threshold period.

Thermosanitration: Heat is any form utilized killing the unwanted microbes from the top soil and below is termed as thermosanitration. It may be flame, simple burning of the refuse, steam smoke or solar heat.

Therapeutants: Fungicide which is capable of eradicating a fungus after it has cause infection and thereby curing the plant.

Thallus: It is a relatively plant body devoid of stem, root and leaves.

Transposon: A segment of chromosomal DNA that can move around (transpose) in the genome and integrate at different sites on the chromosomes.

Toxin: Toxins are extremely poisonous substances and are effective at very low concentrations. Toxins do not affect the structural integrity of the tissues; it affects the metabolism of the plants. Toxins act directly on the protoplast of the cells.

Tissue: Mass of cell or group of cell of similar structure.

Transcription: Copying of a gene into RNA. Also copying of a viral RNA into a complimentary RNA.

Translation: Copying of messenger RNA into protein.

Translocation: Transfer of nutrients or virus through the plant.

Transmission: The spread of a virus or other pathogen from one plant to another.

Transpiration: The loss of water by the areal parts of the plant in the form of invisible water vapour.

Transposable element: A segment of chromosomal DNA that can move around in the genome and integrate at different sites on the chromosomes.

Tumor: An uncontrolled overgrowth of tissues (hyperplasia and hypertrophy of invaded tissues).

Tylosis: An overgrowth of the protoplast of a parenchyma cell into an adjacent xylem vessel or tracheid.

Transfer RNA (t-RNA): The RNA that moves amino acids to the ribosome to be placed in the order prescribed by the messenger RNA.

Transduction: A bacterial virus (phage or bacteriophage) transfer genetic material from the one bacterial cell to other, attacked by them. The attacked cell is not destroyed due to infection by the phage.

Transformation: The change of a cell through uptake and expression of additional genetic material; or bacterial cell absorbed the genetic material in their own cells.

Totipotency: Inherent capacity or potential of a plant cell to regenerate of a whole plant.

Tissue culture: Cultivation in vitro of all plant parts whether a single cell, a protoplast, a tissue or organ under aseptic condition (growth and maintenance of plant tissue or organ).

Transgenic plant: Transgenic plants are plants that have been genetically engineered. It is a modern breeding approach that uses recombinant DNA techniques to create plants with new characteristics *i.e.* transfer of resistance gene into agronomical suitable genotypes or transfer of resistance gene into plant cells or organs or transfer of foreign genes to plant cells or organ. This transformation has been achieved at level of protoplast fusion or cell in many plant species. They are identified as a class of genetically modified organism (GMO).

Tyndalization: Repeated pasteurization until complete sterilization results.

Toxicity: The capacity of a compound to produce injury.

Teliospore: A sexual, thick walled resting spore (part of the basidial apparatus) in which karyogamy takes place (rusts and smuts).

Telium: The fruiting body of binucleate cells in which teliospores are formed.

Trichogyne: The receptive neck of the ascogonium, which is often long and hair-like.

Teleomorph: Sexual or perfect state of a fungus.

Uredium: The asexual fruiting structure of rust fungi in which urediospores are produced.

Urediospore: A dikaryotic (nucleate) spore of the rust fungi.

Uridinium: The sorus of rust fungi producing urediniospores or urediospore (nucleate cell).

Variability: The property or ability of an organism to change its characteristics from one generation to other.

Variant: When a progeny of the pathogen exhibited a characteristic that is different from those present in the parental individuals, it is called a variants.

Virus: Viruses are entities (submicroscopic obligate parasite) whose genome is a mode up of nucleic acid either DNA or RNA and coat protein, which reproduce inside living cells.

Viroids: It is a just like virus but differing from shell which it has devoid off. or

It is a necked nucleic acid, small, circular single stranded, low molecular weight infectious ribonucleic acid (RNA) which is able to cause the disease in the plants and replicate themselves.

Virion: A technical term for the virus, or the complete virus particles.

Virusoid: A sub-viral (extra small) circular RNA component of some isometric RNA viruses. It is a just like viroid except in smallness (extra small).

Virulence: The degree of pathogenicity of a given pathogen, or the disease producing power of any microorganisms or pathogen.

Virulent: Capable of causing a severe disease; strongly pathogenic, highly aggressive or highly pathogenic, with a strong capacity of causing disease deadly very poisonous or harmful.

Vertical resistance (differential or specific resistance): Complete resistance to some races of a pathogen but not to others.

Viral genome: The minimum genetic material of virus required for successful infection and maintenance under field conditions.

Virosis: Infection by a virus.

Viability: The capacity of seed to germinate.

Variety: A strain released for commercial cultivation or in botany a sub-division of a species.

Vector: An organism or agent that can transfer genetic material from one organism called donor or another organism called recipient; or an organism able to transport and transmit a pathogen, especially an inset, nematode *etc.* able to transmit a virus.

Vein banding: A symptom of virus diseases in which the tissues along the veins are darker green than the tissue between the veins.

Vein clearing: A symptom of virus diseases in which the tissues along the veins are lighter green or yellow and the rest of the lamina is normal or dark green.

Vein: A strand of vascular tissue, especially in a leaf.

Vivotoxin: A substance produced internally in an infected host or by the pathogen within it, responsible for some or all of the harmful changes produced in the course of the disease, c. f. pathotoxin.

Warm blooded animal: Those animals which tolerate or resist any changes of environmental condition.

Wilt: The diseases are characterized by drying of the entire plant. The leaves and other green or succulent parts lose their turgidity, to become flaccid and drop; or generalized loss of turgidity and dropping of leaves or shoots.

Wetting agent: A materials which are used to ensure that these will be no layer of air between a solid and liquid; it reduce the surface tension of particles. *e.g.* polyethylene oxide, flour are good wetting agent.

Wettable powder (w.p.): Fungicide formulation composed of active ingredient mixed with fine inert particles (dust) and a wetting agent that mixes readily with water and forms a suspension for spraying.

Witches broom: Clustering of branches of plants resulting in broom-like.

White blister: Characteristic pustular fructification of *Albugo* spp

Yellows: A plant disease characterized by yellowing and stunting of the host plant.

Yeast: A fungus, which reproduces by budding.

Yellow mosaic: A mosaic with pronounced chlorosis of the light areas of the mosaic pattern.

Zoospore: A motile asexually produced spore.

Zoosporangium: A sporangium that contains or produced zoospores.

Zygospore: A sexually resting spore formed by fusion of two gametangia in the Zygomycetes.

Zygote: A diploid cell resulting from the union of two haploid cell or gametes.

References

Abdel M.H, Yamamoto N,Otawa K, Tada C and Nakai Y (2010)Isolation of bacteriumlike substances producing bacteria from finished cattle manure compost and activity evaluation against some food borne pathogenic and spoilage bacteria.*Journal of General and Applied Microbiology*,**56**(2): 151-161.

Adejumo T.O, Ikotun T and Florini D.A (2001)Identification and survival of the causal organism of leaf smut disease of cowpea in Nigeria. *Mycopathologia*,**150**(2): 85-90.

Adejumo T.O, Ikotun Tand Florini D.A (2000)False smut of cowpea is caused by *Protomycopsisphaseoli*(Ramak and Subraman.) in *Nigeria*. Abstracts of paper and poster presented at the *Second World*cowpea *Research Conference*, Accra, Ghana, 3-9 September, 2005.

Agrios G.N(1997) Plant pathology.*4th edition*, *Academic Press*., London, (UK).

Akai S (1951) Studies on the pathological anatomy of fungus gall of plants. Mem.*Coll. Agri. Kyoto Univ*, **58:** 1-60.

AliM.N (2011)Sustainable Agriculture with Low Cost Technologies(SALoCT). School of Agriculture and Rural Development,Ramakrishna Mission Vivekananda University, Belur Math, West Bengal pp. 47.

Alison M.G, and Peter R.F (1999) Insulin releasing and insulin like activity of the

Amine, R., Sarker, B.C., Adhikari, S.K., Sultana, S. and Zubair, T. (2013). Effect of some botanical extracts and cow's urine on *Sclerotium rolfsii*, causal agent of foot and root rot of Betel vine. *The International Journal of Engineering and Sciences* (IJES) Vol.**2** Issue 9 Pages- 77-82.

Anandaraj M (2016) Emerging technologies for sustainable seed spices production. National seminar "New Dimensional Approaches for enhancement of seed spices productivity and profitabltity under era of climate change" ICAR (NRCSS) AJjmer, 2-3 February2016, lec.-**2:** 18-20.

Angua, J.F., Gardner, P.A., Kirkegaard, J.A. and Desmarchelier, J.M. (1994). Biofumigation: Isothiocyanates released from *Brassica* roots inhibit growth of the take-all fungus. Plant Soil, 162: 107-112.

AnjumTehmina, Fatima Sabin, and Amjad Sana (2012) Physiological changes in wheat during development of loose smut. *Tropical Plant Pathology,* **37** (2): 102-107.

Anonymous (1986). Annual Report of Agricultural Research Station. Ananharajupet, A.P. Agri. Univ.

Anonymous (1991). Package of practice. Turmeric. Source: National Research Center for Spices. *Spice India*. **4**: 2-5

Anonymous (1999). Organic farming News letter, **5**(2): 11-15.

Anonymous (2005). Food and Agriculture Organization Production Year Book, Published by Economic and Social Department, FAO, Rome.

Anonymous (2008). USDA nutrient data base from Wikipedia. The free Encyclopedia.

Anonymous (2011). All India Co-ordinate Research Project on Spices, Annual Report. 2012.

Anonymous (2012). All India Co-ordinate Research Project on Spices, Annual Report. 2012.

Anonymous (2013). General information on bio-fumigation. http/www. Infonet-bio-vision.org/default/ct/237/7/20/2013 pp. 1-2.

Anonymous (2014) Annual report of National Horticulture Board New Delhi, Pp.16-17.

Anonymous (2014) Annual Report, National Horticulture Board Darabase 2013-14.

Anonymous: Area and production of vegetable crops, 2012-13. *National Horticulture Board* pp.29.(2013).

Anoop, K. and Suseela Bhai, R. (2014). Evaluation of antagonistic potential of indigenous *Trichoderma* isolates against *Pythium aphanidermatum* (edson) fitz. causing rhizome rot in turmeric (*Curcuma longa* l.). *Journal of science,* **4**(2) 99-105.

Anti-mutagenicity of coriander (*Coriandrum sativum*) juice on the mutagenesis produced by plant metabolites of aromatic amines.*J. Toxicol. Lett,* **153:** 283-292.

Ashlesha and Sugha S.K (2008)*In vitro* antifungal bio efficacy of microbes present in panchgavya against major soil borne pathogens.*Plant Disease Research*,**23**(1): 68-74.

Ashworth, S. and Harrisson (1983). Evaluation of mulches for use in the home garden. *Hort. Sci.*, 18: 180-182.

Atwal, A.S. and Dhaliwall, G.S: Agricultural pests of south Asia and their management. *Kalayani Publishers, New Delhi, India*, 505 (2007).

Awasthi P.B, Mishra U.S and Pande B.N (1978) A new record of *Protomycesmacrosporus* Unger On*Foeniculumvulgare*L. *Curr. Sci.*,**47**: 832-824.

Bacigalova K (2004) Contribution to the knowledge of the *Protomycetaceae*in Slovakia.*ThaiszisJ. Bot.*, **14:** 1-22.

Bacigalova K (2008) *Protomyces buerenianus* (*Protomycetaceae*) a new species for Slovakia.*Biologia*,**63/1:** 40-43.

BacigalovaKamila, PiatekMarcin and WolkowyckiMarek (2005) *Protomyces cirsii-oleracei*(fungi, protomycetales) A new species for Poland. *Polish Botanical Journal*,**50**(1): 77–82.

Bal SS, Singh J and Dhanju KC (1995). Genetics of resistance to mosaic and leaf curl viruses in chilli (*Capsicum annuum* L.). *Indian J. Virol.* **11** (1): 77-79.

Bala, K., G. Robideau, A.W.A.M. de Cock, Z.G. Abad, A.M. Lodhi, S. Shahzad, A. Ghaffar, M.D. Coffey and C.A. Lévesque. (2010). *Phytopythium* Abad, de Cock, Bala, Robideau, Lodhi and Lévesque, *gen. nov. Persoonia*, 24: 136-137.

Balakrishnan, P. (1997). Bio-ecology of rhizome rot of pathogen(s) of ginger and management. Ph.D. thesis, University of Calicut, Calicut, 178.

Banerjee MK and Kalloo G (1987). Sources and inheritance of resistance to leaf curl virus in *Lycopersicon*. *Theor. Appl. Genet.* **73**: 707-710.

Bansal, R.D., Khatri, H.L. and Sharma, O.P. (1985). Growth , sporulation and spore germination of *Colletotrichum capsici* as influenced by tobacco mosaic virus in *Capsicum annuum* L. *Phytopath.* **113**: 36-42,

Bansal, RD., Khatri, H.L. and Sharma, O.P. (1978). Interaction between tobacco mosaic virus and *Collerotrichun capsici* in *Capsicum annuum* L. *Phytopathol.* **17**: 210-21.

Basak A.Band Lee M.W (2005) Efficacy of cow dung in controlling root rot and *Fusarium*wilt of cucumber.*Indian Journal of Plant Pathology*,**23**(1 and 2): 81-84.

Basak, A.B., Lee, M.W. and Lee, T.S.(2002). Inhibitive activity of cow urine and cow dung against *Sclerotinia sclertiorum* of cucumber.*Corean Societyof Mycobiology;* **30 (3)**: 175-179.

Beltagi H.S. E.l, Ahmed O.K andDesoukyW. E.l(2011) Effect of low doses ãirradiationon oxidative stress and secondary metabolites production of rosemary (*RosmarinusofficinalisL.*) callus culture.*RadiatPhys.Chem*, **80**(9): 968-976.

BerucksichtigungihrerEntwicklungsgeschichte and Biologie.*Beitr.Kryptogamenfl. Schweiz*, **5(1):** 1-95.

Bezharuah, S. (2000). Staple Strains. India Today (November): 62-63.

Bhai, S. R., Sasikumar, B. and A. Kumar. (2013). Evaluation of ginger germplasm for resistance to soft rot caused by *Pythium myriotylum*. *Indian Phytopath.* **66**(1): 93-95.

Bhardwaj L. N and Shrestha S. M(1985) Efficacy of fungicide application

Bhat S, Kaushal P, Kaur M and Sharma H.K (2014) Coriander (*Coriandrum sativum* L.): Processing nutritional and functional aspects. *African Journal of Plant Science*, **8**(1)25-23.

Bhaudauria, H.(2002). Cow urine-A magical therapy. Vishwa Ayurveda Parishad. *Int J.Cow Sci.* **1:** 32-36.

Borah, R. K., and Bordoloi, D. K: Influence of planting time on the incidence of leaf curl virus disease and whitefly population on tomato. *Indian Jour. Virol.* **14**(1): 71-73 (1998).

Brayford, D. (1993). The identification of fungi. International Mycological Institute, Kew, Surrey, England. p-27.

Broadbent L and Cooper AJ. 1964. The influence oftomato mosaic virus on root growth and the annuals pattern of fruit production. Ann Appl Biol, 54: 23-30.

Brown, P.D. and Morra, M.J. (1997). Control of soil-borne plant pests using glucosinolates containing plants. *Advances in Agronomy* 61, 167-231.

Brownlee H.E,Mceuen A.R, Hedger J and Scott I.M (1990)Antifungal effects of cocoa tannin on the witches' broom pathogen *Crinipellis perniciosa*. *Physiological and Molecular Plant Pathology*,**36:** 39-48.

Brunt AA and Kenten RH. 1971. Pepper veinal mottlevirus, a new member of the potato Y group from pepers (*Capsicum annuum* L. and *C. frutescens* L.) in Ghana. Ann Appl Biol, 69: 215-43.

Brunt AA, Crabtree K, Dallwitz MJ, Gibbs AJ, Watson Land Zurcher EJ. 1996. Plant Viruses Online: Descriptions and Lists from the VIDE Database. Version: 20th August 1996. URL http: //biology.anu.edu.au/Groups/ MES/vide/Davis RI, Thomas JE, McMichael LA, Dietzgen RG,Callaghan B, James AP, Gunua TG and Rahamma S. 2002. Plant virus surveys on the island of New Guinea and adjacent regions of northern Australia. Austra Pl Path, 31: 385- 390.

Buren G (1922) WeitereUntersuchungenuber die Entwicklungsgeschichte und Biologieder*Protomycetaceen.Beitr.Kryptogamenfl.Schweiz*, **5 (3):** 1-94

Butlar, E.J. (1907). An account of the genus *Pythium* and some chytridiaceae. *Mem. Dept. Agric. India* (Bot, Sci.), **1**(5): 70.

Butler E. J (1918) Fungi and disease in plants.*Thacker spink and co., Calcutta*, pp. 547.

Butler, E.J. and G.R. Bisby, 1931. The fungi of India p. 6, 153. *Sci. Monogrol.* ICAR, New Delhi. (6): 153.

Chadha Sanjay,Rameshwar, Ashlesha, Saini J.P and Paul Y.S (2012) *Vedic Krishi*: Sustainable livelihood option for small and marginal farmers. *Indian Journal of Traditional Knowledge*, **11**(3): 480-486.

Chadha, K.L. (2001). Handbook of horticulture Directorate of information and publications of agriculture. Indian Council of Agricultural Research. New Delhi. p 882.

Chandler, W.A. (1958). Control of bacterial spot and ripe rot of Pimineto pepper PI Dis Reptr. **42**: 652455.

Chandler, W.A. and Dempsey, A.N. (1963). Disinfectant treatments for freshly harvested pepper seed. *PI Dis. Report*. **47**: 325-327.

Chandrasekaran, A. and Rajappan, K. (2002). Effect of plant extracts, antagonists and chemicals (individual and combined) on foliar anthracnose and pod blight of soybean. *J. Mycol. Pl. Patho.* **32** (1), 25-27.

Chandrasekaran, A., Narasimhan, V. and Rajappan, K. (2000). Integrated management of anthracnose and pod blight of soybean. *Ann. Pl. Prot. Sci.* **8** (2), 103-105.

Changes in mustard due to *Altemaria* leaf blight. *Indian Phytopath.*,**43:** 64-69.

Charest M.H, Beauchamp C.J and Antoun H(2005) Effects of the humic substances of deinking paper sludge on the antagonism between two compost bacteria and *Pythiumultimum. FEMS Microbiology Ecology* **52**(3): 219-227.

Charron, C.S. and Sams, C.E. (1999). Inhibition of *Pythium ultimum* and *Rhizoctonia solani* by Shredded Leaves of *Brassica* Species. *Journal of the American Society for Horticultural Science,* **124**(5): 462-467.

Chattopadhya, S. B. (1997). Diseases of plants Yielding Drugs, Dyes and Spices, New Delhi: Indian Council of Agric. Res.

Chattopadhyay, S.B. (1967). Disease of plants yielding drugs, Dyes and Spices, New Delhi, *Indian Council of Agric. Res.*, pp: 345

Chet, I. and Inbar, J. (1994). Biological control of fungal pathogens. *Applied Biochemistry and Biotechnology*, 48: 37-43.

Chong J, Baltz R, Schmitt C, Beffa R and Saindrenan P (2002)Downregulation of pathogen-responsive tobacco UDP-Glc: phenylpropanoids glucosyl transferase reduces scopoletin glucoside accumulation, enhances oxidative stress, and weakness virus resistance. *PlantCell*, **14:** 1093-1107.

Chowdhury, E.K., Hasan, M.M., Mustarin, K., Hasan, M.S. and Fancy, R. (2009). Efficacy of different fungicides in controlling rhizome rot of ginger. *J. Agrofor. Environ.*, **3**(1): 179-181.

Chowdhury, S. (1957). Studies on the development and control of fruit rot of chillies. *Indian Phytopalh.* **10**: 55-62.

Chung Y.R, Kim C.H, Hwang I and Chun J(2000) *Paenibacillus koreensis* sp. *nov.*, a new species that produces an iturin like antifungal compound. *International Journal of Systematic and Evolutionary Microbiology*,**50**(3): 1495-1500.

Chung, W.C., J.W. Huang, H.C. Huang and Jen, J.F. (2002). Effect of ground *Brassica* seed meal on control of *Rhizoctonia* damping-off of cabbage. *J. Plant Pathol.*, 24: 211-218.

Cohen, S., Kern, J., Harpaz, I., and Ben J., R: Epidemiology studies of the tomato leaf curl virus (TYLCV) in Jordan Valley, Israel. *Phytoparasitica* **16**(3): 259-270 (1992).

Cortes-Eslava J, Gomez-Arroyo S, and Villalobos-Pietrini R (2004)

Crover, R.K. and Bansal, R.D. (1970). Seed-borne nature of Colletotrichum capsici in chilli seeds and its control by seed dressing fungicides. *Indian Phytopath.* **23**: 664-668.

Czaczyk K,Trojanowska K and Stachowiak B(2000) Antifungal activity of *Bacillus* sp. isolated from compost. *Folia Microbiology*,**45**(3): 552.

Czech-Kozlowska M and KrzywanskiZ (1984) Phenolic compounds and the polyphenol oxidase and peroxidase activity in callus tissue culture-pathogen combination of red rasberry and Didymellaapplanata (Niessl.) Sacc.*Phytopath. Z*,**109**: 176-182.

DabbasM.R, Singh D.P. and PrakashH.G (2009) Management of Stemgall of Coriander through IDM Practice. *International Journal of Plant Protection*, **2**(2): 55-257.

Daiho, L., Upadhyay, D.N. (2004). Screening of ginger genotypes from North-Eastern States against rhizome rot of ginger. National Agricultural Technology Project Final Report (2001-2004), Nagaland University, Nagaland. 14.

Dake, G.N. and Edison, S. (1988). Survey for disease incidence in major ginger growing areas of Kerala during 1984 and 1985. *J. Plantn. Crops* 16: 55-57.

Darughe F, Barzegar M, and Sahari M.A (2012) Antioxidant andAnti-

Das, K.,Tiwari, R.K.S. and Shrivastava, D.K. (2010). Techniques for evaluation of medicinal plants as antimicrobial agent. Current methods and future trends. *Journal of Medicinal Plants research;* **4** (2): 104-111.

Dator, V.V. and Mayee, C.D. (1981). *Phytopathometry,* Marathwad Agricultural University, Parabhani. pp. 95.

De, B. K., Saha, A. K., and. Nath, P.S: Influence Of meterological factor on the buildup of white fly (*Bemasia tabaci*) vector population and incidence of chilli leaf curl virus disease in the plains of west Bengal. *J. Mycopathol. Res* **43** (2): 195-201(2005).

De, R.K., Mukhopadhyay, A.N. (1994). Biological control of tomato damping off with *Gliocladium virens*. *J. Biological control,* **8**(1): 34-40.

Deadman, M., Al-Hasani, H, Al-Sa'di, A. (2006). Solarization and biofumigation reduce *Pythium aphanidermatum* induced damping off and enhance vegetative growth of greenhouse cucumber in Oman. *J. Pl. patho.,* **88**(3): 335-337.

Dhankar S, Kaur R, Ruhil S, Balhara M, Dhankhar S, and Chhillar A.K, Dick, M.W. (1990). *Keys to Pythium*. Department of Botany, School of Plant Sciences, University of Reading, Reading, U.K. Dickson, B.T. (1925). Colletotrichum v Vrrmicularia. Mycologia. **17**: 213-21 7.

Dimri B. K, Khan, M. N. A. and Narayan M. R (1976) Some promising selection of Bulgaria coriander (*Coriandrum sativum* L.) for seed and essential oil with a note on cultivation and distillation of oil. *Indian Perfumer,***20:** 13-21.

Dixit D, Srivastava NK, Kumar R, Sharma S (2002). Cultivar Variation in Yield, Metabolism Translocation and Partitioning of $14CO^2$ Assimilated Photosynthate into Essential Oil and Curcurmin of Turmeric (*Curcuma longa* L.). *J Plant Biol* **29**: 65-70. 4.

Dogra S. (2006)Anti-fungal potential of panchgavya against some soil borne pathogens.M.Sc. Thesis, Department of Plant Pathology, CSK Himachal Pradesh KrishiVishvavidyalaya, Palampur, India.p 76.

Dohroo, N. P., Kansal, S., Preeti Mehta and Neha Ahluwalia (2012). Evaluation of eco-friendly disease management practices against soft rot of ginger caused by *Pythium aphanidermatum*. *Plant Disease Research* (Ludhiana), **27**(1): 1-5.

Dohroo, N.P., Korla and B.N. and Rattan, R.S. (1988). Effect of chemical seed treatment on pre-emergence rot of ginger. *Proc. Nat. Sem. Chillies Ginger and Turmeric.* Hyderabad (A.P.): 153-155.

Draper, N. R., and Smith, H: Applied Regression Analysis (3rd edition). *New York: Wiley* (1981).

Dubey, N.K., Shukla,R., Kumar, A., Singh, P. and Prakash, B. (2010). Prospects of botanical pesticides in sustainable agriculture, *Curr. Sci. India.* **98:** 479-480.

Dunne, C.P., Dell, B. and Hardy, G.E.S. (2003). The effect of biofumigants on the vegetative growth of five *Phytophthora* species *in vitro*. *Acta Hort.* (602): 45-51.

Durairaj, V. (1956). Growth of *Colletotrichum capsici* in pure culture. *J Indian Bio. Sci.* **35**: 409413.

Eidi M, Eidi A, Saeidi A, Molanaei S, Sadeghipour A, Bahar M, Bahar

Ekbote, S.D.(2005). Management of chili fruit rot caused by *Colltotrichum capsici. J.Mycol. Pl. Path.* **35**(1), 183.

Fageria, M.S., Choudhary, B.R. and Dhaka, R.S. (2006). Vegetable crops production technology, Vol. II, Kalyani Publisher, New Delhi, pp. 223-227.

Fageria, MS; Choudhary, BR and Dhaka, RS (2003) Vegetable Crops: Production Technology. Kalyani Publishers, New Delhi-110002.

Fayzalla, E.A., E. El-Barougy and M.M. El-Rayes, (2009). Control of soil-borne pathogenic fungi of soybean by bio-fumigation with mustard seed meal. *J. Applied Sci., 9: 2272-2279.*

Fisher R. A and Yates (1968) Statistical method for research worker Oliver and Boyd Ltd. *Edin burgh and London*, pp. 10.

Fisher, R. A. and Yates (1968). Statistical method for research worker Oliver and Boyd Ltd. Edin burgh and London. pp. 10.

Frankel E (1995) Nutritional benefits of flavonoids. *International conference on food factors: Chemistry and cancer prevention*, Hamamatsu, *Japan Abstracts*, C**62**.

Friedman P and Rohde R.A (1977) Phenol levels in leaves of tomato cultivars infected with *Pratylenchuspenetrans. J. Nematol.*,**9:** 257-259.

Fulzele, S.V., Satturwar, P.M. and Dorle, a.K. (2001). Immunostimulant activity of Cow's urine. *Journal of Immunology* and *Pathology.* **3**(2): 87-88.

Galindo A, Jeronimo C, Spaans E, and Weil M (2007) An introduction to modern agriculture. *Tierra Tropical*,**3**(1): 91-96

Gamliel, A., Stapleton, J. J. (1993). Characterization of antifungal volatile compounds evolved from solarized soil amended with cabbage residues. *Phytopathology* 83: 899-905.

Garcia, I., Lora, M.J., Cruz, J.D., Benitez, T., Llobell, A. and Pintor-Tora J.A. (1994). Cloning and characterization of a chitinase CHIT42 cDNA from the mycoparasitic fungus *Trichoderma harzianum. Curr Genet*, 27: 83-89.

GaumanE.A and Dodge C.W (1928)Comparative morphology of fungi.*Mc Grow Hill Book Co. Inc.*: pp-701.

Geyter E.D, Lambert E, Geelen D, Smagghe G (2007) Advances of saponins as natural insecticides. *Pest Technology,* **1**(2): 96-105.

Ghosh, R. and Purkayastha, R.P. (2003). Molecular diagnosis and induced systemic protection against rhizome rot disease of ginger caused by *Pythium aphanidermatum, Curr Sci* 85,1782.

Ghosh, R., Datta, M. and Purkayastha, R.P. (2006). *P. aphanisdermatum* induced disease resistance in ginger. *Indian J. Exp. Biol.* 44: 68-72.

Gjaerum H. (1964)*Protomycetes, Protomycopsis* and *Taphridiumin* Norway. *NyttMagasin for Botanik,* **12:** 19–28.

Goel A. K, Kumar S. and Tayal M. S (1983) Biochemical alteration induced in *Coriander sativum* L. by *Protomyces macrosporus* Unger. *Indian Botinical Reporter*, **2:** 62-64.

Gogoi R, Singh D.Vand Srivastava K.D (2001) Phenols asa biochemical basis of resistance in wheat against Karnal bunt. *Plant Pathology*, **50:** 470-476.

Gomathi, V. (2001). Studies on the antifungal effects of plant extracts on *Collerotrichum* spp. infecting *Capsicum annum* L. Ph.D. thesis, Department of Biological Sciences, Pondicherry University, Pondicheny, India.158

Gopal M, Gupta A, Palaniswami C, Dhanapal R and Thomas G.V (2010) Coconut leaf vermiwash: a bio-liquid from coconut leaf vermin compost for improving the crop production capacities of soil. *Current Science,***98**(9): 1202-121.

Gorawar, M.M., Hedge, Y.R. and Kulkarni, S. (2006) Biology and management of spot of turmeric caused by *Colletotrichum capsici, Journal of Plant Disease Sciencc;* **1**(2): 156-158.

Gorawar, M.M., Hedge, Y.R. and Srikant Kulkarni (2005). Chemical control of leaf spot of turmeric caused by *Colletotrichum capsici. Annals of Agri. Bio. Res.* **10**: 2, 159-162.

Green S. K. and J. S. Kim. 1991. Characteristics and control of viruses infecting peppers: a literature review. Asian Vegetable Research and Development Center. Technical Bulletin No. 18. 60 p.

Grewal, A.S, and Grover, R.K. (1974). Changes in the amino acids, Carbohydrates and capsaicin content of red fruits of chilies infected with *Colletotrichum piperanum. Indian Phytopath.* **27**: 603404.

Gupta J. S (1954) Disease appraisal of stem gall of *Coriandrum sativum* L. *Indian phytopath*, **7**(1): 53-60.

Gupta J. S (1958) Laboratory assay of certain fungicides, antibiotics and sulpha drugs on the germination of *Protomyces macrosporus* Unger.*Agra Unive. Jour. Res.* (Sci.), **7:** 105-112.

Gupta J. S (1962) Pathological anatomy of the floral parts and fruits of coriander affected with *Protomyces macrosporus* Unger *Agra Univ. Jour. Res.*, **11:** 307-311.

Gupta J. S and Sinha S (1963a) Biometrical grouping in *Protomyces macrosporus* Unger causing stem gall of Coriander.*Proc. Ind. Acad. Sci.,* **33**(4): 507-510.

Gupta J. S and Sinha S (1963b)Some therapeutic control trials on the stem gall disease of coriander. Bull. *Indian Phytopath.* Soc., **1:** 75-78.

Gupta J. S and Sinha S (1964) Variation in pathogenicity of *Protomyces macrosporus* Unger. *Proc. Natn. Acad. Sci.*, **34**(3): 241-244.

Gupta J. S and Sinha S (1973) Varietal field trial in control of stem gall disease of coriander.*Indian Phytopath.*,**26**(2): 337-340.

Gupta P.P, Gupta S.K, Kaushik C.D. and YadavaT.P. (1985) Biochemical changes in leaf surface washings of groundnut due to tikka disease *Cercosporidium personatum. Indian Phytopath.*, **38:** 339-340.

Gupta R. N (1973) Longevity of chlamydospores of coriander gall fungus. *Indian phytopath.*,**26**(3): 581-582.

Gupta R. N. (1975) Mineral matter content in coriander leaves and fruits as influenced by stem gall disease. *Indian phytopath.*,**28:** 136-137.

Gupta R. N (1976) Impact of nitrogen, phosphorus and manganese on the stem gall of coriander. *ActaBotanicaIndica*, **40**(1): 30-35.

Gupta S.K, Gupta P.P, Yadava T.P and Kaushik C.D (1990) Metabolic

Gupta, S.K. and Tripathi, S.C. (2011). Fungitoxic activity of *Solanum torvum* against *Fusarium sacchari. Plant Protection Science*, **47**(3): 83-91.

Gupta, V. K., Sharma, R., Singh, S., Jindal, J and Dilawari, V. K: Efficiency of *Bemisia tabaci* (Gennadius) populations from different plant host for acquisition and transmission of cotton leaf curl virus. *Indian journal of Biotech.* **9**: 271-275(2010).

Harborne J.B (1973) Phytochemical Methods.*1st ed. Chapman and Hall, London*, 273.

Harsh Deep Yadav, Poonam Yadav, Lokendra Yadav and Prabhat Kumar Yadav (2013). Efficacy of chilli mottle virus disease on root constraints of *Capsicum annuum* L. at Agra, India. *World J. of Applied Sci. and Res.* 5759.

Hasnat, M., Rajib Bose, Hossain, M.A. Anam, M.M., Kabir, H. (2014). Effect of chemicals, bio-agent plants extract and soil amendments in controlling rhizome rot of ginger. *International J. Natural and Social Sci.*, 1: 01-11.

Haware M. P and Pavgi M. S (1976) Host Range of Two *Protomycopsis* Species from India. Faculty of Agriculture, Banaras Hindu University, India, 272-274.

Haware M.P and Pavgi M.S (1976) Chlamydospore germination in *Protomycopsis*species. Faculty of Agriculture, Banaras Hindu University, India.*Mycopathologia*,**59**(2): 105-111.

Haware M.P, and Pavgi M.S (1971) Perpetuation of *Protomycopsis* Species Causing Purple Leaf' Spot of Legumes in India.*Ann. Phytopath. Soc. Japan*, **37:** 242-248.

Henig-Severa N, Eshelb A., and Neemana G (2000) Regulation of the germination of Aleppo pine (*Pinushalepensis*) by nitrate, ammonium and gibberellin, and its role in post-fire forest regeneration.*Physiologiaplantarum*.**108**: 390-397.

Hidayat HS, Opriana E, Manzila I and Sujiprihati S. 2012. Occurrence of *Chili Veinal Mottle Virus* (ChiVMV) in Indonesia and response of chili *World Journal of Applied Sciences and Research, 2013, 3(1): 5759*

Horowitz, A. R: Population dynamics of *Bemisia tabaci*(Gennadius): with special emphasis on cotton field., **17**: 37-47(1986).

Horowitz, A.R., Podoler, H., Gerling, D: Life Table Analysis of the Tobaco Whitefly *Bemisia tabaci* (Gennadius) in Cotton Fields in Israel. *Acta Oecologica Oecologia Applicata* **5**(3), 221-233 *Agric. Eco. Environ* (1984).

Howell CR. (1991). Biological control of *Pythium* damping-off of cotton with seed coating preparations of *Gliocladium virens*. *Phytopathology*, 81, 738-741.

Hudge, B.V., Ghugul, S.A. (2009) Evaluation of phytoextracts against *Colletotrichum capsici* causing leaf spot of turmeric, *Annals of Plant Physiology;* **23**(2): 276-277..3.

Isaac S. (1992)Fungal life-style. In: *Fungal plant-Interactions*, Chapman and Hall, London, England, p. 1.

Isabelle Louis, Chew, A. and Lim, G. (1988). Influence of spore density and extracellular conidial matrix on spore germination in *Colletotrichum capsici*. *Trans. Br. Mycol. Soc.* 91: 694497.

J. Talukdar, A. K. Saikiaand P. Borah (2015). Survey and detection of the diseases of Bhut Jolokia (Capsicum chinense Jacq.) in Assam. *Journal Crop and Weed.* **11**: 186-192.

Jagtap, G.P. (2012). Control of *Colletotrichum trancatum* causing anthracnose/ pod blight of soybean by aqueous leaf extracts, bio-control agents and fungicides. *Scientific Journal of Agriculture*. **1**(2) 39-52.

Jagtap, G.P., Mali, A.K. and Utpal D. (2013). Bio efficacy of fungicides, Bio-control agents and botanicals against leaf spot of turmeric incited by *Colletotrichum capsici*. *African Journal of Microbiology Research;* Vol. **7**(18): 1865-1873.

Jain, N.K., Gupta V.B., Garg R. and Silawat (2010) N. Efficacy of cow urine therapy on various cancer patients in Mandsaur District, India - A survey. *Int J Green Pharm;* **4:** 29-35.

JaindaikSavita, Thakur preeti and Kumar Vikas (2015) Efficacy of cow urine as plant growth enhancer and antifungal agent. *Advances in Agriculture*, **7**: ID 620368.

Janardhanan M and ThoppilJ. E (2004) Herb and spice essential oil.*Discovery publishing House*, New Delhi, 110002: 40-42.

Jayalakshmi, C., Durairaj, P., Seetharaman, K. and Sivaprakasam, K. (1998). Bio-control of fruit rot and die-back of chili using antagonistic microorganisms. *Indian Phytopath.* **51** (2), 180- 183.

Jayanta, T.C.R. (2003). Differential reactions to Colletotrichum leaf spot and genetic variability in some cultivars of turmeric. *Annals of Plant Protection Sciences.* **11**: 300-303.

Jeumkya, H. and Byungkook, H. (1998). Influence of inoculum density, wetness duration, plant age and inoculation method and cultivar resistance on infection of pepper plants by *Colletotrichum capsici. Plant Dis.* **82**: 1079-1083. 012

Johnston, A. and Booth, C. (eds)(1983). Plant Pathologist's Pocket-Book, 2nd edn. Kew, Surrey, U.K.: Commonwealth Mycological Institute.

Jones, D.R: Plant viruses transmitted by whiteflies. *European J.Plant Pathol.,* **109**: 195–219 (2003).

Joseph B and Sankarganesh P (2011) Antifungal efficacy of panchagavya. *International Journal of Pharm Tech Research,* **3:** 585-588.

Joseph Philip and Nair, P.C.S. (1981). Field reaction of turmeric types to important pests and diseases. *Indian Cocoa Arecanut and Spices Journal.* **4**: 107-109.

Joseph, L. and Sivaprasad. (2000). Biological control of rhizome rots of ginger. *In*: Proceedings of Indian Phytopathological Soceity Golden Jubilee and International Conference on Integrated Plant Disease Management for Sustainable Agriculture, Vol. I. at New Delhi, from November, 11-15, 1997, pp339- 340.

Joshi, C.G. and Wangikar, P.D. (1978). Efficacy of fungicides against *Colletotrichum capsici. Indian Phytopath.* **31**: 222-223.

Joshi, D. and Tripathi, H.S. (2002). Cultural, biological and chemical control of anthracnose of urdbean. J. *Mycol. Pl. Path.* **32** (1), 52-55.

Joshi, L. K. and Sharma, N. D. (1982). Diseases of ginger and turmeric. In: Nair M K, Premkumar T, Ravindran P N and Sarma Y R (Eds.). Proceedings, National Seminar on Ginger and Turmeric, Calicut (pp. 104.117). Central Plantation Crops Research Institute, Kasaragod.

Joshi, L.K. and Sharma, N.D. (1980). Diseases of Ginger and Turmeric, Proceedings of the National Seminar on Ginger and Turmeric held at Calicut. April 8-9. pp 104-109.

Kai H, Ueda T and Sakaguchi M. (1990)Antimicrobial activity of bark-compost extracts. *Soil Biology and Biochemistry,***22**(7): 983-986.

Kallupurackal, J.A. and Ravindran, P.N. (2002). Turmeric: hints for cultivation. *Spice India* **15**: 6-1 1.

Kambar, Y., Vivek, M. N., Manas, M., Kekuda, P.T.R., Nawaz, N.A.S. (2013). Inhibitory effect of cow urine against *Colletotrichum capsici* isolated from anthracnose of chili (*Capsicum annum* L.). *Science Technology and Arts Research Journal.* **2(4)**: Article in press.

Kar, A.K. and Mahapatra, H.S. (1981). New host records of Colierotrichum species from India. *Indian Phytopath.* **34**: 219-221

Katan, J., Greenberger, A., Aton, H. and Grinstein, A. (1976). Solar heating by polythene mulching for the control of diseases caused by soil borne pathogens. *Phytopathology*, 66: 683-688.

Kaur, M., Sharma, O.P. and Sharma, P.N. (2006). In vitro effect of Trichoderma species on C. capsici causing fruit rot of chili. *Indian Phytopath.* **59** (2), 243-245.

Kaur, S., Singh, J. and Sooch, B.S. (1985). Natural occurrence of different spp, of Collero~rirh~rm attacking Capsicum annum L. *Indian Phytopulh.* **38**: 755.

Kavyashree, R. (2009). An efficient *in-vitro* protocol for clonal multiplication of Ginger var.varada. *Indian J. of Biotech.*, 8: 228-231.

Keshwal R.L and Khatri R.K (1998) Reaction of some high yielding varieties of coriander to powdery mildew.*J, of Myco.and Pl. Patho,* **28**(1): 58-59.

Khaing N.S, Khai A.A, Aye K.S and Khaing O.Z (2008)Maintenance of beneficialmicrobes in vermiwash carrier system and its application in crop. *GMSARN International,***14**(3): 12-16.

Khatso, Kevimeo and N. Tiameren Ao. (2013). Biocontrol of Rhizome Rot Disease of Ginger (*Zingiber officinale* Rosc.). *International Journal of Bio-resource and Stress Management*, **4**(2): 317-321.

Kim, J.W. and Park, E.W. (1999). Occurrence and Pathogenicity of *Pythium species* Isolated from Leaf Blight Symptoms of Turfgrasses at Golf Courses in Korea. *Plant Pathol. J.*, **15**(2): 112-118.

Kirkegaard, J.A., Wong, P.T.W., Desmarchelier, J.M. and Sarwar, M. (2006). Suppression of soil-borne cereal pathogens and inhibition of wheat germination by mustard seed. Proceedings of the 13th Australian Agronomy Conference, September 10-15, 2006, Australian Society of Agronomy.

Kirkgaard, J. A., Mathiessen, N.P., Wong, T.T.W., Mead, A., Sarwar, M., Smith, B.J. (1999). Exploring the bio-fumigation potential of Brassicas in farming systems. Proceedings of 10th International conference on rapeseed, Canberra, Australia.

Kirtikar K.Rand Basu B.D (1956)Indian Medicinal Plants.*2nd Edition.*Lalit Mohan Basu,Allahabad, Uttar Pradesh, India, **III:** 2045–2048.

Koche Mina; kothikar, R.B. and Patil, C.U. (2009). Fungicides and botanicals for managing foliar diseases of turmeric. *J.Pl. Dis. Sci.,* **4**(1): 68-71.

Konate, K., Kiendrebeogo, M., Ouattara, M.B., Souza, A., Lamienmeda, A., Nongasida, Y., Barro, N., Millobo-rasolodimby, J. and Nacoulma, O.G. (2001). Antibacterial potential of aqueous acetone extract from five medicinal plant used traditionally to treat infectious diseases in Burkina faso. *Current Research of Biological Sciences.***3** (5): 435-442.

Kulkarni (2011). Etiology, epidemiology and integrated management of rhizome rot of ginger in Karnataka. *Indian Phytopath.*, **64**(2): 113-119.

Kulkarni, S., Sagar, S.D. and Lokesha. N.M. (2005b). Integration of soil solarization in the management of rhizome rot of ginger. Paper presented in 'National symposium on modern concepts in plant disease management and their global thrust' and I.P.S. Annual meet (South Zone), Annamalai University, Annamalainagar, T.N. from 29-30 Dec. p.24.

Kulkarni, S., Sagar, S.D., Hegde, Y.R. and Lokesha. N.M. (2004). In vitro screening of botanicals against *Pythium aphanidermatum* and *Fusarium solani* causing rhizome rot of ginger. Paper presented in symposium on Biology, Biotechnology. Epidemiology and management of plant diseases and IPS annual meeting (South zone) held at Gulbarga University, Gulbarga, Karnataka.

Kumar S, Kumar S, Singh M, Singh AK, and Rai M (2006). Identification of sources and inheritance of Pep LCV in *Capsicum. Sci. Hort.* 110**:** 359-361.

Kumar Sanjay (2008) Screening of germplasm and genetics of resistance against leaf curl virur of *Capsicum.* Ph.D. Thesis, Udai Pratap Autonomous College, Varanasi, India.

Kumar, A., Mohanty, A.K., Borah, T.R., and Singh, A.K. (2012). Bio-mulching in organic ginger production under rain-fed conditions in NE hills region. *Indian J. Traditional Knowledge;* **11**(3): 487-492.

Kumar, A., Reeja, S. T., Bhai, R. S. and Shiva, K. N. (2008). Distribution of *Pythium myriotylum* Drechsler causing soft rot of ginger. *J. Spices and Aromatic Crops.* **17**(1): 5-10.

Kumar, R., Rai, N. and Lakpale, N. (1999). *Orissa Jour. of Hort.* 27: 100-102.

KumarGagan, Yadav S.K, Patel J.S, SarkarA, and Awasthi L.P (2014) Management of Stem Gall Disease in Coriander Using *Pseudomonas* and *Trichoderma* (Bio agents) and Fungicides. *Journal of pure and applied microbiology,***8:** (6) p. Proof.

Kushalappa A. C and Ludwig A (1982) Calculation of apparent infection rate in plant diseases: Development of a method to correct for host growth. *Phytopathology,***72:** 1373-1377.

Kushalappa, A. C. and Ludwig, A. (1982). Calculation of apparent infection rate in plant diseases: Development of a method to correct for host growth. *Phytopathology,* 72: 1373-1377.

Kushwah, S.S., Dwivedi, Y.C. and Jain, P.K. (2013) Effect of mulching materials on growth and yield attributes and enhancing farm income through ginger cultivation under rainfed rice based production system. *Vegetable Science,* **40**(1): 96-98.

Kuvalekara A, RedkaraA, Gandheb K and HarsulkaraA (2011) Peroxidase and polyphenol oxidase activities in compatible host-pathogen interaction in

Jasminum officinale and *Uromyces hobsoni*: Insights into susceptibility of host. *New Zealand Journal of Botany*, **49**(3): 351-359.

Lakra B.S (1990) Effect of depth and amount of inoculum and time of inoculation of chlamydospores of *Protomyces macrosporus* in infectivity and severity of coriander. *Plant Disease Research*, **5**(1): 118.

Lakra B.S (1999) Assessment of losses due to stemgall in coriander.*Plant Disease Research*, **14**(1): 85-87.

Lakra B.S (2001) Estimation of chlamydospores of *Protomyces macrosporus* in different infected plant parts of coriander. *Plant Disease Research*, **16**(1): 9495.

LakraB.S (1993) Effect of depth and amount of inoculum and time of inoculation of chlamydospores of *Protomyces macrosporus* in infectivity and severity of coriander.*Plant Disease Research*, **5**(1): 118.

LakraB.S (2000) Management of stem gall of coriander (*Coriandrum sativum*) incited by *Protomyces macrosporus. Indian Journal of Agricultural Sciences*,**70**(5): 338340.

Larkin, R.P. and Griffins, T.S. (2007). Control of soilborne potato diseases using *Brassica* green manures. *Crop Prot.*, 26: 1067-1077.

Latif, M. A., and Akhter, N: Population Dynamics of Whitefly on Cultivated Crops and its Management. International Journal of Bio-resource and Stress Management 4(4): 576-581(2013).

Lewis Y.S (1984) Spices and Herbs in food, Industry.*Food Trade Press Oxpuigton, UK*, pp. 200.

Li, Y.; Mao, L. G.; Yan, D. D.; Liu, X. M.; Ma, T. T.; Shen, J.; Liu, P. F.; Li, Z.; Wang, Q. X.; Ouyang, C. B.;Guo, M. X. (2014). First report in China of soft rot of ginger caused by *Pythium aphanidermatum*. *Plant Disease*, **98**(7): 10-11.

Lily V.G and RamadasanA (1979) Changes in phenolic content of coconut leafin relation to the development of leaf rot. *Indian Phytopath*, **32:** 112-113.

Lin, L.T., S.S. Chang and L.S. Leu (1971). Soft rot of ginger. *Pl. Prot. Bull. Taiwan*, 12: 54-67.

Lodhi, A.M., Ali Khanzada, M., Shahzad, S. and Ghaffar, A. (2013). Prevalence of *Pythium aphanidermatum* in agro-ecosystem of Sindh province of Pakistan. *Pak. J. Bot.*, **45**(2): 635-642.

Lokesh, M. S., Patil, S. V., Gurumurthy, S. B., Palakshappa, M. G., Anandaraj, M. (2012). Solarization and antagonistic organisms for management of rhizome rot of ginger in Karnataka. *International J. Plant Protec.*, **5**(2): 195-200.

Lorena T, Calamassi R, Mori B, Mugnai L, and Surico G (2001) *Phaeomoniella chlamydospora*_grapevine interaction: histochemical reactions to fungal infection. *Phytopathologia Mediterrranea*, **40: 400**-406.

M. Ashfaq , S. Iqbal, T. Mukhtar and H. Shah (2014). Screening for resistance to cucumber mosaic cucumovirus in chilli pepper. *The Journal of Animal and Plant Sciences*, **24**(3): 791-795.

M. H. Rashid, K. M. Khalequzzaman, M. S. Alam, S. A. Uddin And S. K. Green (2007). Screening of different sweet pepper lines against cucumber mosaic virus and chili veinal mottle virus. *Int. J. Sustain. Crop Prod.* **2**(3): 1-4.

Maheshwari D.K andChaturvedi S.N (1981) Histochemical localization of DNA and histones in the hypertrophied parts of *Coriandrum sativum* due to stem gall disease. *Indian Phytopath*, **34**(4): 488-491.

Maheswari, N. U. Sirchabai, T. P. (2011) Effect of *Trichoderma* species on *Pythium aphanidermatum* causing rhizome rot of turmeric. *Biosciences, Biotechnology Research Asia*, **8**(2): 723-728.

Makkar H.P.S, Norvsambuu T. L, khavatsere S and Becker K (2009) Plant secondary metabolites in some medicinal plants of Mongolia used for enhancing animal health and production. *Tropicultura*, **27**(3): 159-167.

Manandhar, J.B., Hadman, C.L. and Wang, T.C. (1995a). Conidial germination and appressorial formation of *Colletorrichum capsici* and *Colletotrichum gloeosporioides* isolates from pepper. *Plant Dis*. **79**: 361-366.

Manandhar, J.B., Hadman, C.L. and Wang, T.C. (1995b). Semiselective medium for Colletotrichum gloeosporioides and occurrence of three *Colletotrichum* spp. on pepper plants. *Plant Dis*. 79: 376-379.

Mansangwa, J.I.G., Aveling, P.A.S. and Kritzinger (2013). Screening of plant extract for antifungal against *Colletotrichum spp*. of common bean (*Phaseolus vulgaris* L.) and cowpea (*Vigna anguilata* (L.) Walp. *Journal* of *Agricultural Science*.**151:** 482-491.

Masoko, P., Picarid, J. and Eloff, J.N. (2005).Antifungal activities of six South African *Terminalia species* (Combreta ceae). Journal of Ethonopharmacology **99:** 301-308.

Mathai, P.J., Dubey, G.S., Peter, K.V., Saklani, U.D., Singh, N.P., 1977. Pant C-1 and Pant C-2—two new promising selections of chilli *Capsicum annuum* L. South Indian Hort. 25, 123–125.

MathivananR, Edwin, S.C,Viswanathan K, and Chandrasekaran D (2006) Chemical Microbial composition and antibacterial activity of modified panchagavya.*International Journal of Cow Science*, **2**(2): (retrieved 12.02.10 from *Indian Journals.com*).

Mathur S.B (1962) Soil P^H in relation to stem gall disease of coriander caused by *Protomyces macrosporus* Unger. *Indian Phytopatho.*,**15:** 75-76.

Mathur S.B and Narula P.N (1963) Effect of certain fungicides, antibiotics and synthetic phytohormonesthe germination of chlamydospores of *Protomyces macrosporus* Unger causing stem gall of coriander.*Proc. Ind. Acad. Sci. Sec.*, **33**(4): 615-617.

Mathur, Kusum; Ram, D., J. Poonia and Lodha., B.C. (2002). Integration of soil solarization and pesticides for management of rhizome rot of ginger. *Indian Phytopath.* **55**(3): 345-347.

Matthiessen, J.N. and Kirkegaard, J.A. (1998). Biofumigation for managing soil-borne pests- progress, pitfalls and prospects. Proceedings of the 6th Australasian Applied Entomology Research Conference (Eds Zalucki M, Drew R and White G). Volume 1: 364-372, University of Queensland.

Mayton, H.S., Olivier, C., Vaughn, S.F. and Loria, R. (1996). Correlation of fungicidal activity of *Brassica* species with allyl-isotiocyanate production in macerated leaf tissue. *Phytopathology* 86: 267-271.

Mc Donald, B.A. and Linde, C. (2002). Pathogen population genetics, evolutionary potential, and durable resistance. Annual Review of Sclerotium rolfsi, *Phytopathology*, 71: 1099-1003.

Mc Rae (1911). Efficacy of fungicides for seed treatment against rhizome rot of ginger. *Indian Journal of Plant Pathology*, **19**(1 and 2): 39-43.

McRae. W. 1908. *Rep. Agric Sta. E. Bengal and Assam, p.* 30 (1911). Soft rot of ginger in the Rangpur district. *East Bengal Agric. J.* India, 6: 139-146.

Meena, R. Panti, V. and Arora, D. K: An epidemic of chilli leaf curls disease in Rajasthan. *J. Phytol. Res.* **19**(2): 335-336 (2006).

*Meloidogynegraminicola*infected rice plants (*Oryza sativa* L.).*International Journal of Advanced Research*, **1**(6)71-76.

Memane SA, Joi MB, and Kale PN (1987). Screening of chilli cultivars against leaf curl complex. *Current Research Reporter*. Mahatma Phule Agricultural University **3**(1): 98-99.

Memane SA, Joi MB, and Kale PN (1987). Screening of chilli cultivars against leaf curl complex. *Current Research Reporter*. Mahatma Phule Agricultural University **3**(1): 98-99.

Mishra R. S and Pandey V. P (2013) Effect of bio agents and fungicides on the management of stem gall of coriander.Presented in National symposium on "Emerging pollutant and pathogen due to climate change challenges and risk reduction", Pp-82.

Mishra R.S and Pandey V.P (2015) Evaluation of varieties of coriander (*Coriandrum sativam* L.) for resistance to stem gall disease and seed yield. *Curent advances in Agricultural Sciences*, **7**(2): 151-153.

Mishra R.S and Pandey V.P(2015) Screening of *Coriandrum sativum* accessions for stem gall disease tolerance and seed yield. J. *Plant Disease Science,* **10** (2): 256-259.

Mishra R.S, Ramveer and Pandey V.P (2015) Antifungal effect of cow urine extracts against *Colletpotrichum capsici* causing leaf spot disease of turmeric, *African Journal of Biotechnology,* **14**(20): XXX-XX, XX-XXX.

Mishra R.Sand KumarSandeep (2016)Viability of coriander seeds from *Protomyces macrosporus* infected plants to spread of stem gall of coriander. CAAS: **26**(02): in press.

Mishra, A.C. and Pandey, V.K. (2009) Effect of mulching on plant growth and rhizome yield in ginger (*Zinziber officinale* Rosc.) in rainfed plateaus of Jharkhand. *International J. Pl. Sci;* **4**(2): 385-387.

Mishra, R.S. and Pandey, V.P. (2014). Management of leaf spot of turmeric caused by *Colletotrichum capsici* through fungicides. "National symposium on plant pathology in genomic era and 66th Annual meeting of IPS". 26th to 28th may 2014 Raipur pp.22.

Mishra, R.S. and Pandey, V.P. (2014). Management of soft rot of ginger through Bio-Fumigation, Fungicides and bio-agents. *J. Pl. Dis. Sci.* **9**(1): 128-130.

Mishra, R.S. and Pandey, V.P. (2015). Management of *Colletotrichum* leaf spot of turmeric through fungicides. *J. Spices and Aromatics Crops;* **24**(2): 153-156.

Mishra, R.S. and Sandeep Kumar (2015). Evaluation of plant extract, bio-agents and fungicides against *Fusarium udum* causing pigeon wilt. *J. Pl. Dis. Sci.* **10**(1) 92-98.

Mishra, R.S., Pandey, V.P. and Ankit Tiwari (2015). Screening of genotypes and effect of fungicides against leaf blotch of turmeric. *J. Pl. Dis. Sci.* **10**(1) 99-102.

Misra, A.P. and Dutta, K.K. (1963). Studies in anthracnose fungi, 111. A comparative study of two isolates of *Colletotrichum capsici* (Syd.) Butler and Risby. *J Indian Bio Sci.* **42**: 74-84.

Misra, A.P. and Mahmood, M. (1960a). Factors affecting the growth of *Collerotrichum capsici. Indian Phytopath.* **13**: 12-1 7.

Misra, A.P. and Mahmood, M. (1960b). Effects of carbon and nitrogen nutrition on growth and sporulation of *Colletorrichum capsici* (Syd.) Butler and Bisby. *J Indian Bio Sci* **39**: 314-321.

Mohamed A.A, Khalil A.A, Beltagi H.E.SE.l.(2010) Antioxidant and antimicrobial properties of kaff Maryam (*Anastatica hierochuntica*) and doum palm (*Hyphaenethebaica*) Grasas Y Aceites,**61**(1): 67–75.

Mohan B (2008) Evaluation of organic growth promoters on yield of fruits crops in *India. J. Organ. Syst.*, **3**(1): 23-36.

Mohanty K.C, Dash L, and Swain S.C (2001) Effect *of Meloidogyneincognita* on nodulation and root biochemistry in French bean.*Ann. Plant Protec. Sci.*,**9:** 81-86.

Mohanty, I., Senapati, M.R., Jena, D. and Palai, S. (2014). Diversified uses of cow urine. *International Journal of Pharmacy and Pharmaceuticals Science. Vol.* **6** Issue 3: 45-48.

Mordue, J.E.M. (1971). *Glomerella cingulata.*C.M.I. Description of Pathogenic fungi and Bacteria. No-315.

Muhammad S and Amusa N.A (2003) *In vitro* inhibition of growth of some seedling blight inducing pathogens by compost inhibiting microbes.*African Journal of Biotechnology,***2**(6): 161-164.

Mukhopadhyay A. N (1968) Studies on stem gall of coriander caused by *Protomyces macrosporus*Unger. Ph. D. Thesis, B. H. U. Varanasi,Pp-152.

Mukhopadhyay A.N and Pavgi M.S (1964) Chlamydospore germination of artificial culture of *Protomyces macrosporus* Unger.*Experimentia,***20**(11): 619.

Müller K O,and Börger H (1940) Experimentelleuntersuchungenuber die *Phytophthoraresistenzder*kartoffel. *Arb.Biol. Reichsasnstalt Landw Forstw Berlin,* **23:** 189–231.

Mundkur B.B (1949) Fungi and Plant Disease. *Mc. Milan and co. Ltd., London.* sp. 107-109.

Mundkur B. B (1959) Fungi and plant diseases. *Nacmillan and Co. Ltd., London,* pp. 246.

Muruganandam, V., Suryanarayanan, R.S. and Venkataehalam, S. (1987). Formation of infection structure of *Colletotrichum capsici* as influenced by nitrogen sources. *Curr Sci.* **13**: 674475.

Nabavi S.M, Ebrahimzadeh M.A, Nabavi S.F, Hamidinia A, Bekhradnia A.R (2008) Determination of antioxidant activity, phenol and flavonoids content of *Parrotiapersica*Mey. *Pharmacologyonline,* **2:** 560-567.

Nageshwar Rao, T.G., Reddy, B.N. and Sasikumar, B. (1993). Turmeric: production constraints of turmeric in Andhra pradesh. *Spice India.* **6**: 8-11.

Naidu, P.H. (1988). Spices Newsl. Spices Board, Cochin, June pp. 10-13.

Naik N., and Sreenivasa M.N (2009) Influence of bacteria isolated from Nakasone A. K, Bettiole W, and Souze R.M.D (1999) The effect of water extracts of organic matter on plant pathogens. *Summa Phytopathologica,* **25:** 330-35.

Naqvi S. A. M. H. (1986) Varietal screening of coriander against stem gall in relation to disease intensityand crop loss. *Ind. Jour. of Myco. and plant patho.,***16**(3): 270-276.

Narain, A. and Das, D.C. (1970). Toxin production during pathogenesis of *Colletotrichum capsici* causing anthracnose of chili. *Indian Phytopath.* **23**: 484-490.

Narain, A. and Panigrahi, C. (1971). Efficacy of some fungicidal compounds to control *Colletotrichum capsici in vitro* and *in vivo. Indian Phytopath* **24**: 593- 596.

NarasimhaRao, S., Ravindra, K.K. and Anandraj, M. (2012). Management of leaf spot of turmeric (*Curcuma longa* L.) incited by *Colletotrichum capsici* through fungicides.*journal of Spices and Aromatic Crops*.Vol. **21**(2): 151-154.

Narasimhudu, Y, and Balasubramanian, K.A. (2001b). Laboratory evaluation of fungicides against *Colletotrichum capsici* causing leaf spot of turmeric. *Pesrology*. **25**: 31-32.

Narasimhudu, Y. and Balasubramanian, K.A. (2001a). Effect of fungicides on growth and vigour of turmeric. *Pesrology*. **25**: 20-21.

Narula P. N and Joshi M. G (1963) Differential response of coriander strains to stem gall disease. *Science and Culture*, **29**(4): 206.

Natarajan K (2002)Panchagavya: A manual, Other India Press, Mapusa, Goa, India, pp. 33.

Ndoumou D.O, dzomo T .G. N and jocgoue P.F. D (1996) Changes in Carbohydrate, Amino Acid and PhenolContents in Cocoa Pods from Three Clones after Infection with *Phytophthora megakarya* Bra. and Grif. Ann. of Botany, **77:** 153-158.

Nduaguc, C., Ekefan, E.J. and Nwankiti, A.O.(2008). Effect of some crude plant extract on growth of *Colletotrichum capsici* (Synd.) Butler and Bisby, causal agent of pepper anthracnose, *Journal of Applied Bio-Sciences;***6** (2): 184-190.

Neeraj and Vermashilpi (2010) Quantitative changes in sugar and phenolic contents of *Brassica* leaves induced by *Alternaria brassicae* infection. *International Journal of Plant Protection*, **3**(1): 28-30.

Nene Y. L, Siddiqui I.A and Kharbanda P.D. (1966) Control of stem gall of coriander by fungicide. *Mycopath. Mycol. Appl.*, **29**(1-2): 142-144.

Nene, Y.L. and Thaplial, P.N. (1993). Evaluation of fungicides in; Fungicides in plant disease control (2011 new ed.) *Oxford, IBM Publishing Co.*, New Delhi, 331 pp.

Nikam B.C, Kulkami, S.N and Pawar P.K (1959) Coriander its diseases. *Indian Forming*, Nov.12-13.

Nishida H and SugiyamaJ (1994) Archiascomycetes: detection of a new lineage within theAscomycota. *Mycoscience,* **35:** 361-366.

Nishida H, Ando K, Ando Y, Hirata A and Sugiyama J. (1995) Mixiaosmundae: transfer from the Ascomycota to the Basidiomycota based on evidence from molecules and morphology. *Can. J. Bot.*,**73:** 660-666.

Nyam M.A, Wonang D.L, Akueshi C.O (2009) Phytochemical screening and antimicrobialstudies on *Canariumschiveinfurthii* Linn ("Atili") fruits and oil. Nigerian Journal of Botany,**22** (2): 247-253.

Okigbo, R.N. and Nameka, I.A. (2005). Control of Yam tuber with leaf extracts of *Xylopia aethiopica* and *Zingiber officinale. Afr. J. Biotechnol.* **4**(8): 804-807.

Okigbo,R.N. and Ogbonnaya, U.O.(2006). Antifungal effect of two tropical plants leaf extracts (*Ossimum gratinnimum* and *Aframomum meleguata*) on post harvest yam (*Diascortea* Spp.) rot. *African Journal of Biotechnology*, **5**: 727-731.

Ong CA, Vorghese G and Ting WP. 1986. The effect of chilli veinal mottle virus on yield of chilli (*Capsicum annuum* L.), MARDI Res Bull, 8: 74- 8.

Osbourn A.E (1996) *Plant Cell*,**8:** 1821–1831.

Osbourn A.E (2003) Saponins in cereals. *Phytochemistry.* **62:** 1-4.

Osbourn A.E (1996) Preformed antimicrobial compounds and plant defense against fungal attack. *Plant Cel*, **8:** 1821-1831.

Palarpawar, M.Y. and Ghrude, V.R. (1989). Saprophytic colonization of leaf sheath of turmeric by *Colletotrichum capsici* and *Colletotrichum curcumae. Indian J. of Mycol. PI Patho* **19**: 202-203.

Palarpawar, M.Y. and Ghrude, V.R. (1997). Influence of different nitrogen sources on growth and sporulation of *Colletotrichum capsici* and (*Colletotrichum curcumae*) *J. of Mycol. and PI. Pathol.* **27**: 227-228.

Panchagavya on seed germination and seed vigour in wheat.

Pandey R.N and Dange S.R.S (1998) Diseases of coriander and fennel- a review. *Agricultural Review Karnal*, **19**(2): 120-125.

Papadopoulou K, Melton R.E, Leggett M, Daniels M.J. and Osbourn A.E (1999) Compromised disease resistance in saponin-deficient plants. *P. Natl. Acad. Sci. USA.*,**96** 22): 12923-12928.

Papen H, Gabler A,ZumbuschE and Rennenberg H. (2002) Chemo litho autotrophic nitrifiers in the phyllosphere of a Spruce ecosystem receiving high nitrogen input. *Curr. Microbiol.*,**44:** 56-60.

Parihar Singh Pankaj (2012) Changes in metabolites of *Brassica juncea* (Indian mustard) during progressive infection of *Alternaria Fbrassicae. Nature and Science*, **10**(3): 39-42.

Park, M. (1941). Report on the work of the mycological division, Ceylon Administration Reports, *Rep. Dir. Agric. Ceylon*, 1939: D-20-D-22.

Parmasto E. and Parmasto I. (1987) Variation of basidiospores in the Hymenomycetes and its significance to their taxonomy.*Bibl. Mycol*, **115:** 1–168

Parveen, T. and Sharma, K. (2014). Management of "Soft Rot" of Ginger by Botanicals. *Int. J. of Pharm. Life Sci.* **5**(4): 3478-3484.

Pathak M.L and Kumar a (2003) "Cow praising and importance of panchagavya as medicine," *Sachitra Ayurveda*, **5**): 56-59.

Pathak, M.I. and Kumar, A. (2003). Cow prasing and importance of panch gavya as medicine. *Sachitra Ayurveda;* **5**: 56-59.

Pathak, M.I. and Kumar, A. (2003). Go-mutra destructive study. *Sachitra Ayurveda.* **7**: 81-84.

Patil, A., Laddha, A., Lunge, A., Paikrao, H.P. and Mahure, S. (2012). In vitro antagonistic properties of selected *Trichoderma species* against tomato root rot causing *Pythium species. International J. Sci., Environ and Tech,* **1**(4): 302–315.

Patil, M.J., Ukey, S.P. and Raut, B.T. (2001). Evaluation of fungisides and botanicals for the management of early blight (*Alternaria solani)* of tomato . *PKV Research Journal* , **25:** 49-51.

Paul, B., K. Bala, L. Belbahri, G. Calmin, E. Sanchez- Hernandez and F. Lefort. (2006). A new species of *Pythium* with ornamented oogonia: morphology, taxonomy, ITS region of its rDNA, and its comparison with related species. *FEMS Microbiology Letters,* 254: 317-323.

Pavgi M. S and Mukhopadhyay A. N (1969-a) Transmission of *Protomyces macrosporus* Unger, the incitant of stem gall of coriander. *Ann. Phyto. Patho. Soc. Japan,* **35**(4): 265-270.

Pavgi M. S and Mukhopadhyay A. N (1969-b) Viability of seed in coriander infected by *Protomyces macrosporus* Unger. *Ann. Phyto. Patho. Soc. Japan,* **35**(4): 271-274.

Pavgi M.S And Mukhopadhyay A.N (1970) Cytology of chlamydospore germination in *Protomyces macrosporus* Unger. Faculty of Agriculture, Banaras University, India.*Cytologia,* **35:** 359-367.

Pavgi M.S and Mukhopadhyay A.N (1972) Development of coriander fruit infected by *Protomyces macrosporus* Unger, **37:** 619-627.

Pavgi M.S and Mukhopadhyay A.N (1973) Cytology of Chlamydospore Development in *Protomyces macrosporus* Unger. . Faculty of Agriculture, Banaras Hindu University, India.*Cytologia,* **38:** 467-477.

Paxton J.D (1981) Phytoalexins-a working redefinition.*Phytopathol Z,* **101:** 106-109.

PayalLodha and ParasharAnamika(2007) Phenolic estimation in *Foeniculumvulgare* infected with ramularia blight.*Annals of Plant Protection Sciences,* 15(2): 396-398.

Pinto S.S, Mann M.B, Campos F.S, Franco A.C, Germani J.C and Van Der SST(2010) Preliminary characterization of some *Streptomyces* species isolated from a composting process and their antimicrobial potential.*World Journal of Microbiology andBiotechnology,***26**(10): 1847-1856.

Plotnikova L.Z(1977)Dung infusion as a means for controlling powdery mildew of rose.*ZashchitaRastenii,* **2:** 28.

Popta C.M.L (1899)BeitragzurKenntnis der Hemiasci Flora, **86:** 1-46.

Poudyal, B.K. (2011) *Commercial Ginger Farming in Nepal, Today and Future.* Sajha Prakashan Press, Kathmandu, Nepal.

Prabhu M.J (2009) Panchagavya: low cost organic input for both crops and animals.In: Farmer's Notebook, *The Hindu Online Edition*, **4**[th]June 2009.

Prasad B. K, Singh R. N and Narayan, N (1989-b) Biochemical changes in nitrogen and carbohydrate in coriander infected with *Protomyces macrosporus* Unger. *Indian Phytopath.*,**42**(3): 426-430.

Prasad B.K (1983) Free sugar and amino acid contents of coriander infected with stem gall disease. *Indian Phytopath.*,**36**(1): 150-151.

Prasad B.K, Singh R.N, Shankar U and Krishore A (1989-a) Alteration in in the respiration of coriander infected with *Protomyces macrosporus* Unger causing stem gall. *Indian Phytopath.*,**42**(4): 534-537.

Prashith Kekuda, T.R., Vivek, M.N., Manasa, N., Kambar, Y., Noornawaz, A.S. and Raghvendra, H.L. (2014). Antifungal effect of cow urine extracts of selected plants against *Collrtotrichum capsici* isolated from anthracnose of chili. *International Journal of Scinces.* **7** (3): 142-143.

Price K. R, Johnson, I. T and Fenwick, G. R (1987) *CRC Crit. Rev. Food*

Pring, R.J., Nash, C., Zakaria, M. and Bailey, J.A. (1995). Infection process and host range of *Colletotrichum capsici. Physiol. Mol Pl Parhol.* **46**: 137-152.

Pruthi J.S (1976) Spices and condiments.4th *Ed.*, National Book Trust,

Pruthi, J.S. (1998). Major spices of India Crop management, post-harvest technology. Publ. Publication and Information Div., ICAR, New Delhi.

Pureifull DE, Zitter TA and Hiebert E. 1975.Morphology, host range, and serological relationships of pepper mottle virus.

Purseglove, J.W. (1968) Turmeric. In Spices. Vol. 2. Longman Inc. New York, pp- 532-579.

Rai, Bimla, M.M. Jha, K.L. Ojha and S.B. Sharma (1998). Effect of fungicidal seed treatment on Rhizome rot disease of ginger. *Nat. Sem. Rec. Dev. Spi. Tech.* B.A.C. April, 20-21, 1998 (pp. 43-44).

Raj Abhishek, Kumar Manoj, ToppoPratapJhariya, (2014) Cow dung for ecofriendly and Sustainable productive farming. *Environmental Science,* **3:** 2277-8179.

Raja J and Kurucheve V (1999) Fungicidal activity of buffalo (*Babulusbubalis*) urine: a new record. *Madras Agricultural Journal* **86**(10/12): 614-616.

Rajak, C.R (1983). Nutritional requirements of two species of *Colletotrichum. Acro Boranico Indico.* **11**: 155-160.

Rajan, K.M. and Agnihotri, V.P. (1989). *Pythium* induced rhizome rot of ginger-problems and progress. In: Agnihotri V P,

Rakesh K.N, Dileep N, Syed Junaid, PrashithKekuda T.R, Vinayaka K.S and Noor Nawaz A.S (2013) Inhibitory Effect of Cow Urine Extracts of Selected Plants against Pathogens Causing Rhizome Rot of Ginger. *Science, Technology and Arts Research Journal*, **2**(2): 92-96

Ram, D. and Patil, R.K. (2014). Evaluation of fungicides, antagonists, phytoextracts and cow urine against *Fusarium pallidoroseum* causing fruit rot of tomato. *J.Pl. Dis. Sci.*, **9**(1): 100-104.

Ram, P., Mathur, K. and Lodha, B. C. (1999). Integrated management of rhizome rot of ginger involving biocontrol agents and fungicides. *J. Myco. and Pl. Patho.*, **29**(3): 416-420.

Ramakrishnan, T.S. (1949). The occurrence of *Phytium vaxans* de Bary in South India. *Indian Phytopath*, 2: 27-30.

Ramakrishnan, T.S. (1954). Le-af spot disease of turmeric (Curcumu ir~ngu L) caused by *Colletotrichum capsici* (Syd.) Butler and Bisby. *Indran Phytopath.* **7**: 111-117.

Ramkumar, Suoreche, R., Prabhakar, S. and Pandurangan, M. (2012). Role of antagonistic microbe *Psuedomonas florescense* on *Colletotrichum capsici* infecting Curcuma longa. *J Plant Pathol*. Microb 3: 146. doi: 10.4172/2157-7471.1000146.

Ramteke, P. K. and Kamble, S. S. (2011). Physiological studies in *Fusarium solani* causing rhizome rot of ginger (*Zingiber officinale* Rosc.). *The Bioscan*. **6**(2): 195-197.

Rangaswami, G. and Mahadevan, A (2001). Diseases of crop plants of India. IV Edition, (Eds.) Prentice hall of lndia (P.) Ltd., New Delhi. pp.453-455.

Rao C. P (1972) Anatomical studies on abnormal growth caused by *Protomyces macrosporus* on *Coriandrum sativum* L. *Indian Phytopath.*, **25**(4): 483-496.

Rao, C.H. and Narayana, Y.D. (2005). *In vitro* evaluation of fungicides, plant extracts and bio-control agents against *C. dematum* (Pers. Ex. Fr.) Grove the causal organism of chickpea (*Cicer arietenum* L.) blight. In national symposium on crop disease management in dry land Agril. and 57th Annual meeting IPS, Jan. 12-14, 2005, MAU, Parbhani.

Ravishanker, Kumar, S., Chatterjee, A., Baranwal, D. K. and Solankey, S.S. (2014). Genetic variability for yield and quality traits in ginger (*Zingiber officinale* Roscoe). *The Bioscan*, **8**(4): 1383-1386.

Reddy M. S and Kramer C. L (1975) A taxonomic revision of the Protomycetales. *Mycotaxon*, **3**: 1–50.

Reddy, S.G., Dakshinamurthy, V. and Sarma, S.S. (1963). Note on varietal resistance against leaf spot disease in turmeric. *Andhra Agric J.* **10**: 146-148.

Rezk M.A, Ibrahim, I.K.A and Ibrahim A.A.M (1976) Effect of root-knot nematodes on phenolics contents of barley and wheat.*Nematol. Medit.* **15:** 259-263.

Ricker, A.J. and Ricker, R. S. (1936). Introduction to research on plant diseases. John, S Swift Co., 4.4,USA.

Roberts, R.G. and Snow, J.P. (1990). Morphological and pathological studies of *Colletotrichum capsici* and *C. indicum*. *Mycologia*. **82**: 82-90.

Rohit R.P, Kavindra R.J and Chintan K.M, (2011) Image morphological operation based quality analysis of coriander Seed (*Coriandrumsativum* L.) Emerging Trends in Networks and Computer Communications (ETNCC).*International Conference,* **22-24**: 482-486.

Rojasara, Y.M., Rajani, M.D., and Dhruj, I.U. (2014). Management of groundnut rust through host plant resistance and fungicides. *J.Pl. Dis. Sci.,* **9**(1): 38-42.

Romero, E., Zurera, C., Porras, M., Barrau, C. and Romero, F. (2007). Biofumigant effect of *Brassica* species in the growth of *Phytophthora spp. in vitro.*15th International Reinhardsbrunn Symposium: Modern Fungicides and Antifungal Compounds. Friedrichroda, Germany 6-10 May.

Ross, I.A. (1999). Medicinal Plants of the World. Chemical constituents, traditional and modern medicinal uses. Humana Press.

Sagar, S.D. (2006). Investigations on the etiology, epidemiology and integrated management of rhizome rot complex of ginger and turmeric. Ph.D. Thesis, Department of Plant Pathology, University of Agricultural Sciences, Dharwad-580005

Sagar, S.D., Srikant kulkarni, Hedge, Y. R., Rao, M.S.L. (2007). Management of rhizome rot of ginger by bio agents. *Annals of Agriculture Bio Research,* **12**(1): 57-59.

Sahib N.G, Anwar F, Gilani A.H, Hamid A.A, Saari A and Alkharfy

Sahni, S.S., Chabal, D.S. and Nirmaljit Singh. (1974). *In-vitro* production of toxic metabolite by different species of Colletotrichum. *Indian J Mycol. PI Parhol.* **4**: 222-223.

Sangamaheswara Sarma and Krishna Murthy, D. (1962). Varietal resistance against leaf spot diseases of turmeric. *Andhra Agric J.* **9**: 6165.

Sangeetha V, Thevanathan R (2010a) Effect of TMV strains in a local lesion host. Journal of panchagavya on nitrate assimilation by experimental Ecobiol.,**31:** 28-131.plants,*The journal of American Science,* **6:** 80-86.

Sanger, R.B.S., Katawale, T.R., Sarof, R.K. and Parihar, M.S. (1988). *Farm Sci. Jour.* **3**: 69-71.

Sanjay Kumar, Rajesh Kumar, Sanjeet Kumar, Major Singh, A.B. Rai And Mathura Rai (2009). Reaction of pepper leaf curl virus field resistance of chilli (*capsicum annuum* L.) genotypes under challenged condition. *Veg. Sci.* **36**(2): 230-232.

Sanjutha S, Subramanian S, Rani I. and Maheswari (2008) Integrated Nutrient Management in *Andrographis paniculata. Research Journal of Agriculture and Biological Sciences*, **4**(2): 141-145.

Sarma, Y.R. (1994). Rhizome rot disease of ginger and turmeric. In: Chadha K L and Rethinam P (Eds.) Advances in Horticulture, Vol.10. Plantation and Spices Crops, Part-2. (pp. 1113.1138). Malhotra Publication House, New Delhi.

Sarwar, M., Kirkegaard, J.A., Wong, P.T.W., and Desmarchelier, J.M. (1998). Bio-fumigation potential of brassicas. In vitro toxicity of isothiocynates to soil borne-fungal pathogens. *Plant and Soil*, 201: 103-112.

Saunders, W.H.M.(1982). Effect of cow urine and its major constituents on pasture properties .*NZ J. Agricul. Res.* **25** (1): 61-68.

Saxena M, Saxena J, Nema R, Singh D and Gupta A (2013) Phytochemistry of medicinal plants.*J. Pharm. Phytochem.*, **1**(6): 168-82.

Saxena R.P, Dixit, J, Pandey U. P and Singh V.K (2002) Germplasm screening for stem gall disease of Coriander. Paper presented in National Symposium on integrated management of plant disease of Mid-Eastern India with a cropping system perspective, held at NDUAT, Faizabad, Dec. 5-7-2002.

ScarpariL.M,Meinhardt L.W,Mazzafera,PomellaP.A.Schiavinato A. M, CascardoJ.C.M and PereiraG. A. G (2005) Biochemical changes during the development ofwitches' broom: the most important disease ofcocoa in Brazil caused by *Crinipellisperniciosa.Journal of Experimental Botany*, **56**(413): 865-877.

Schoemaker, I.A.S. Ghini, R. (2001). Bio-fumigation for *Pythium spp.* control (Portguese). *Summa Phytopathologica*; 27(3): 308-312.

Schonbeck F and Schlosser E (1976) in *Physiological Plant Pathology*, eds.Heitefuss R. and Williams, P. H. (Springer, Berlin), pp. 653–678.

Schönbeck F andSchlösser E (1976) Preformed substances as potential protectants. In: Heitefuss R, Williams P H,editors. Physiological plant pathology. Berlin, Germany: *SpringerVerlag* KG, 653-678.

Schrandt, J.K., Davis, R.M. and Nunez, J.J. (1994). Host range and influence of nutrition, temperature and pH on growth of *Pythium violae* from carrot. *Plant Dis.* **4**(78): 335-338.

Senanayake, D. M. J. B., Verma A., and Mandal, B: Virus- Vector Relationship, Host Range, Detection and sequence comparison of chilli leaf curl associated with an epidemic of leaf curl disease of chilli in Jodhpur, India. *J. Phytopathol* **160**: 146-155(2012).

Senapati, G.R. and Sugata Ghose (2005). *Indian Phytopatho.*, 58: 437-439.

Sengupta , D.K., Maity, T.K. and Dasgupta, B. (2009) Effect of mulching on ginger (*Zingiber officinale* Rosc.) in the hilly region of Darjeeling district. *Journal of Crop and Weed,* **5**(1): 203-205.

Shah H, Khalid S and Ahmad I. 2001. Prevalence and distribution of four pepper viruses in Sindh, Punjab and North West Frontier Province. On Line J Biol Sci, 1: 214-217.

Shahare, K.C. and Asthana, R.P. (1962). Rhizome rot of ginger and its control, *Indian Phytopath.,* 15 (1): 78 (RAM) **42**(1): 41, 1963.

ShailajaB,MishraIpsita,GampalaSrihima, SinghVikramJeet and Swathi K (2014) Panchagavya- an eco-friendly insecticide and organic growthPromoter of plants.*International Journal of Advanced Research,* **2:** (11) 22-26.

Shakuntala N.M, Vasudevan S.N, Patil S.B, Doddagoudar S.R,Macha RCMSI Vijay and Kumar A.G (2012) Organic bio priming on seed vigour inducing enzyme in paddy - An alternative to inorganics. *Ecoscan,* **1:** 251-257.

Shalika (2009) Non-Chemical management of stalk rot of cauliflower caused by *Sclerotiniascle rotiorum*(Lib.) de Bary. M Sc Thesis, Department of Plant Pathology, CSK Himachal Pradesh KrishiVishvavidyalaya, Palampur, India, p 48.

Sharma A and Deshpande V.K (2006) Effect of pre-soaking of pigeon pea seeds with organics on seed quality. *Karnataka Journal of Agricultural Sciences,***19:** 396-399.

Sharma OP and Singh J (1985). Reaction of different genotypes of pepper to cucumber mosaic and tobacco mosaic viruses. *Caps. Newsl.* 4: 47.

Sharma, B. R., Dutta, S., Roy, S., Debnath A. and Roy, M. D. (2010). The effect of soil physico-chemical properties on rhizome rot and wilt disease complex incidence of ginger under hill agro-climatic region of West Bengal. *Plant Pathology J.,* **26**(2): 198-202.

Shepherd RJ and Purcifull DE. 1971. Tobacco etch virus CMI/AAB, In: Descriptions of plant viruses, pp: 55. Taufik M, Astuti AP, and HidayatSH. 2005. Survey of *Cucumber mosaic virus* and *Chilli veinal mottle virus* on chilli pepper and screening of resistance of chillipepper genotypes (in Bahasa Indonesia, English abstract). J Agricult, 16: 146-152.

Shivapuri, A., Sharma, O.P.and Jhamaria, S.L.(1997). Fungitoxic properties of plant extracts against pathogenic fungi. *Journal of Mycology and Plant Pathology;* **27**: 29-31.

Siddiquie, M.R., Dharam Sigh and Ashok Gaur (1977). Prevalence of chilli anthracnose hgus on seeds and its effective control. *Seed Res.* **5**: 67-72.

Singh G and Bedi D.S (1976) Phenolic and sugar constituents of gram cultivarsresistant and susceptible to *Operculella padwickii.Indian Phytopath.,***29:** 191-192.

Singh G.R, Chaure N.K and Parihar S. S (2001) Organic Farming for sustainable agriculture *Indian Farmer*, **52:** 12-17.

Singh H.B, Singh A, Tripathi A, Rai S.K, Katiyar R.S, Johri J.K, and Singh S.P. (2003) Evaluation of Indian coriander accessions for resistance against stem gall disease. *Genetic Resources and Crop Evolution*, **50:** 339-343.

Singh K,Vashistha RN, Pandita ML and Batra, BR (1976) Effect of mulching on growth and yield of cucurbits under rainfed conditions. *Haryana J. Hort. Sci.*, **5**(1 and 2): 87-91.

Singh P and Thapliyal P. N (1987) Effect of some soil factors on disease controlling potential of Carbendazim.*Indian Phytopath.*,**40**(2): 204-208.

Singh S. P, Gupta J. S and Sharma A. K (1984) Disease appraisal and crop loss estimates in coriander attacked by *Protomyces macrosporus* Unger in Uttar Pradesh. *Geobios*, **11**(6): 276-278.

Singh SJ (1973). Reaction of chilli Cultivars (*Capsicum sp.*) to mosaic and leaf curl viruses in field conditions. *Indian J. Hort*. **30**(1-2): 444-447.

Singh, A.K. and Edison, S. (2003). Eco Friendly management of leaf spot of turmeric under partial shade. *Indian Phytopathology*.**56**: 4,479-480.S

Singh, A.K., Basandari,A. and Sharma, B.K. (2003). Fungicidal management of leaf spot of turmeric. *Indian Phytopathology*. **56** (1): 119-120.

Singh, A.K., Singh, R.K., Singh, A.K., Singh, V.K., Rawat, S.S., Mehta, K.S., Kumar, A., Gupta, M.K. and Thakur, S. (2014) Bio-mulching for ginger crop management: Traditional ecological knowledge led adoption under rainfed agro-ecosystems. *Indian J. Traditional Knowledge*, **13**(1) 111-122.

Singh, H. K., Shakywar, R. C., Shyam Singh, Singh, A. K. (2012). Evaluation of comparative efficacy of native isolate of *Trichoderma viride* against rhizome rots disease of ginger. *J. Pl. Dis. Sci.*, **7**(1): 22-26.

Singh, R. S. (2002). Introduction to Principles of Plant Pathology. Fourth Edition, Oxford and IBH Publishing Co. Pvt. Ltd., New Delhi. pp. 279-292.

Singh, R.S. (1963). Plant disease Oxfords IBH Publishing Co., New Delhi-1.

Singh, R.S. (2006). Plant Disease. 8th Edition Oxford and IBH Publishing Co. Ltd. New Delhi, 167-168.

Singh, S. Gupta, B. R., Pande, S. P. and Tripathi, D. P: Response of some chilli lines to mosaic and leaf curl diseases, *Indian J. Virol.*, **6**(1-2): 113-115 (1990).

Singh, S. J., Sastry, K. S. and Sastry, K. S. M: Combating leaf curl virus in chilli. *Ind. Hort.*, **24**: 9-10(1979).

Singh, S. J., Sastry, K. S. and Sastry, K. S. M: Efficacy of different insecticides and oil in the control of leaf curl virus in chilli. *Zeit. Pflanzens.*, **86** (5): 253-256 (1979).

Singh, S.K., Rai, B., Kumar, B. (2004). Evaluation of different fungicides in controlling the rhizome rot of ginger under storage and field conditions. *Annals of Agri Bio Research,* **9**(1): 63-65.

Singh, S.P. (1999). Biological Control in India. *Ind J PI Pror.* 126-138

SinghAmitabh, Kumar JaiswalRitesh, MauryaSudarshan, and SinghUdaiPratap (2013) Analysis of phenolic and indole acetic acids in

Sinha R.K, Valani D, Chauhan K and Agarwal S. (2010)Embarking on a second green revolution for sustainable Agriculture by vermin culturebiotechnology using earthworms: Reviving the dreams of Sir Charles Darwin.*Journal of Agricultural Biotechnology and Sustainable Development,***2:** 113-128.

Sinha, B., and Trafdar, J: Epidemiological study of SPVD transmission in West Bengal. *Bioscan.* **9**(3): 1237-1242 (2014).

Sivan A., Y. Elad and I. Chet. (1984). Biological control effects of a new isolate of *Trichoderma harzianum* on *Pythium aphanidermatum.* Phytopathology, 74: 498-501.

Snapp, S. S., O'Neil, K., Kirk, W. W., Bird, G., Quakenbush, B. (2004). Improving short potato-based rotations through organic amendments. Michigan 2004 Potato Research Report. Michigan State University Agricultural Experiment Station, in cooperation with The Michigan Potato Industry Commission 36: 22-27.

Somasundaram, E and Amanullah M.M (2007) Panchagavya on growth and productivity of crops: *A review.Green Farming,* **1**: 22-26.

Sparg S.G, Light M.E and Staden J (2004) Biological activities and distribution of plant saponins.*J Ethnopharmacol,***94:** 219-243.

Sreenivasa M.N and Naik N (2011) Impact of bacteria isolated from organic liquidmanures on seed germination and seedling vigour index of wheat and soybean. *Organic Farming Newsletter,***7:** 3-5.

Srikant Kulkarni (2011) Etiology, epidemiology and integrated management of rhizome rot of ginger in Karnataka. *Indian Phytopath.,* **64**(2): 113-119.

Srimal, R.C. (1997). Turmeric: A brief review of medicinal properties. *Fitoperapia* **68**(6): 483-493.

Srimathi P, Mariappan N., SundaramoorthyL and Paramathma M (2013)

Srivastava H.C (1955-b) Studies on tumour disease of *Coriandrum sativum* L. caused by *Protomyces macrosporus* Unger.*Proc. 42nd Ind. Sci. Cong.,* Part **III**: 218.

Srivastava U, Rai R.A and Agrawal J.M (1971b)Powdery mildew of coriander and its control.*Indian Phytopath.,***24**(3): 437-440.

Srivastava, V.P. and Gupta, J.H. (1977). Fungicidal control of turmeric leaf spot incited by *Taphrina maculans. Ind J Mycol. PI Pathol.* **7**: 76-77.

Stanghellini, M.E. and Russell, J.D. (1973). Germination *In vitro* of *Pythium aphanidermatum* oospores. *Phytopathology,* 63: 133-137.

Stansly, P. A., Schuster, D. J. and Liu, T. X: Apparent parasitism of Bemisia argetifolii (Homoptera: Aleyrodidae) by Aphelinidae (Hymenoptera) on vegetable crops and associated weeds in South Florida. *Biological control,* **9**: 49-57 (1997).

Stephanie, G. Harvey and Carl E. Sams. University of Tennessee (2001). Brassica bio-fumigation increases marketable tomato yield, Department of plant Sciences and Landscape Systems, Knoxville, TN 37901.

Stirling, G. R., Turaganivalu, U., Stirling, A. M., Lomavatu, M. F. and Smith, K.M. (2009). Rhizome rot of ginger (*Zingiber officinale*) caused by *Pythiu m myriotylum* in Fiji and Australia. *Australasian Plant Pathology,* 38, 453–460.

Subhashini, D.V. (2013) Bio-efficacy of plant extracts against major pathogens affecting tobacco nurseries, *Annal Plant Protection Science*: **20**(2): 414-416.

Subrahmaniyan, K., Zhou, W. and Veeramani, P. (2011). Weed control through degradable, herbicidal plastics and organic mulches and their effect on crop growth and yield of winter rapeseed (*Brassica napus*). *Indian J. Agril. Sci.,* **81**(4): 348-352.

Subrahmaniyan, L.S. (1919). A *Pythium* disease of ginger. *Madras Agric. J.,* 38: 225-226.

Sugha S.K (2005) Antifungal potential of panchgavya.*Plant Disease Research* **20:** 156-158.

Suleiman, M.N., Emua, S.A. and Ayodele, S.M. (2011) Growth and physiological studies on root rots fungus of cowpea. *European Journal of Experimental Biology,* **1**(4): 181-188.

Suprapta, D.N. and Khalimi, K. (2009). Efficacy of plant extracts formulations to suppressed stem rot disease on Vanilla seedlings. *J. ISSAAS Vol.* **15** No.2: 34-41.

Swaminathan C, Swaminathan V and Vijay Lakshmi K (2007) Panchgavya boon toorganic farming.*International Book Distributing Co. Lucknow,* P. 332.

Swamy, B.S. and Kulkarni, S. (2003). In vitro evaluation of fungicides and botanicals against *Colletotrichum capsici* (Syd) causing leaf spot of turmeric. *Indian Phytopath.* **56** (3), 339.

Sydow, H. (1913). Beitrage Zur Kenntnis der Pilzflora des sudlichen Osrindiens I *Ann Mycol.* **11**: 329-330.

Tambe V.D, Bhambar R.S (2014) "Estimation of Total Phenol, Tannin, Alkaloid and Flavonoid in *Hibiscus Tiliaceus*Linn. Wood Extracts" *RRJPP,* **2**(*4*)*:* 41-47.

Tarafdar, J. and Saha, N. (2007).Correlation study on population dynamics of ginger soft rot inciting pathogens under different organic amendments, disease incidence and its survival in Darjeeling hill soils. Proceedings of the 13th ISTRC Symposium, pp.165-169.

Tayal K. S, Kumar S, Geol A. K and Maheshwari D. K (1981) Role of IAA, IAA oxidase o-dihydroxy phenols, poly phenol oxidase and peroxidase in stem gall disease of *Coriandrum sativum* L. *Current Science*, **50**(17): 785-786.

Tewari, V.P., Ramanujam, S., 1974. Grow Pusa Jwala, a disease resistant high yielding chilli. *Indian Farming* 24, 20.

Thakur S (2010) Studies on some plant growth promoting traits in bacteria isolated from vermiwash. M Sc Thesis.Department of Microbiology, College of Agriculture, CSK HPKV, Palampur, India, p 86.

Thamburaj, S. (1991). Research on spice crops at TNGDNAU. *Spice India*. **4**: 17-21.

TharmarajK,GaneshP, Suresh KumarR, AnandanA and KolanjinathanK(2011)"A critical review on Panchagavya—a boon plant growth," *International Journal of Pharmaceutical and Biological Archive*, **2**(6): 1611–1614.

Thind, K.S. and Randhawa, H.S. (1957). Studies on the nutrition of (*Colletotrichum capsici* (Syd.) Butler and Bisby), the incitant of die-back of chilli. *Curr Sci* **26**: 17-18.

Thirumalachar M. J and Pavgi M. S (1950) Notes on spore germination and mounting techniques.*Indian Phytopath.*,**3:** 177-178.

Thirumalachar M. J, Pavgi M. S, and Singh R.A (1969) In vitro activity of 3 antifungal antibiotics against *Protomyces macrosporus* Unger, *Tapharinamaculans*and *Corticumsasakii.Hindustan Anti.Biot. Bull.*, **11**(3): 189-190.

ThirumalacharM.J and Narasimhan M. J (1953) Notes on some mycological methods.*Mycologia*, **45:** 461-468.

Thomas, K.M. 1938. Detailed administration Report of the Govt. Mycologist, Madras, 21 pp.

Thomas, K.M. 1940. Detailed administration Report of the Govt. Mycologist, Madras, 30 pp.

Thomas, K.M. 1941. Detailed administration Report of the Govt. Mycologist, Madras, 18 pp.

Tiwary R. S and Agrawal A (2004) Production technology of spices.*1st ed. International Book Distribution co.* Chaman Studio Building, 2nd floor, Charbagh, Lucknow, 2260004 U.P., India, pp-254-271.

Toncer O,Tansi S and Kizil S (1998)The effect of different sowing time on essential oil of coriander in GAP region.*Anadolu.*,**8**(2): 101-105.

Tripathi A.K (2001) Evaluation of coriander cultivars for resistant against stem gall disease and yield potncial.*Crop Research Hisar*, **22**(3): 485-488.

Tripathi A.K (2003) Effect of sowing dates and weather parameters on stem gall disease of coriander. *Indian Phytopath.*,**56**(2): 191-193.

Tripathi A.K and Bartaria A.M (2002) Chemical control of stem gall disease (*Protomyces macrosporus)* of coriander and its economics. *Plant Disease Research*, **17**(1): 119-120.

Tripathi A.K, Pandya R.K, Tiwari R.J and Tripathi M.L (2002) Response of stem gall disease *(Protomyces macrosporus)* and yield of coriander *(Coriandrum sativum)* to soil pH. *Indian J. Agri. Sci.*, **11:** 49-50.

Tubki K (1957) Biological and cultural studies of three species of *Protomyces. mycologia*, **49**: 44-54.

Tuite, J. (1969). Plant pathological methods fungi and bacteria, Burgess Publishing Company, Minneapolis Min. 55

Turner E. M (1953) *J. Exp. Bot.*,**4**: 264- 271.

Usman, N.M., Balakrishnan, P. and Sarma, Y.R. 1996, Biocontrol of rhizome rot of ginger. *Journal of Plantation Crops*, 24: 184-191.

Van der Plaats-Niterink, K. J. (1981). Monograph of the genus *Pythium*. Stud. Mycol. 21: 1-24.

Van Os, G. and L. Lazzeri, (2006). Control of *Rhizoctonia solani* in lily by biofumigation with *Brassica* seed meal. Proceedings of the 2nd International Biofumigation Symposium, June 25-29, University of Idaho, Moscow.

Vasudeva, R. S: Report of the division of mycology and Plant Pathology. *Sci. Rept. Agric Res* 79-89(1954).

Vasumathi R(2001) Influence of organic manures, bio fertilizers and plant density on growth, yield and alkaloid content of Bhumyamalaki (*Phyllanthus amarus* Schum.And Thonn.) M.Sc., (Hort.) Thesis TNAU , Coimbatore-3. India.

Verma, A. K., Basu, D., Nath, P. S., Das, S. and Ghatak, S. S: and Mukhopaghyay, S: Relationship between the population of whitefly, *Bemisia tabaci* Genn. (Homoptera: Aleyrodidae) and the incidence of tomato leaf curl virus disease. *Ind. J. Mycol. Res.* **27** (1): 49-52(1989

Verma, R.K., Singh, Y., Soni, K. K. and Jamalluddin (2005). Solarization of forest nursery soil for elimination of root pathogens and weeds. *Indian J. Trop. Biodiv.*, 13: 81-86.

Vikas and Ratnoo,R.S. (2011). Evalution of fungicides and phytoextract against powdery mildew (*Erysiphe polygoni* de.*)* of fenugreek.Journal of Plant Disease Scinces. Vol-**6** issue 1. Pp-71-73.

Vrang, J., Rasochova, M., Ficker, A. and Dobias, K. (1990). Inoculation of potatoes with micro-organisms under field conditions. 1. Effect of plant growth, yield and physiological properties of tubers in potato and sugerbeet. *Folia Microbiologica*, 35: 326-335.

Wate, S. T., Dhanjode, D.P., Duragkar, M.J. and Tajne, M.R. (2011). Antioxidant potential of cow urine and its fractions: A comparative *study: Inter J .Uni Pharm Life Sciences.***1**(1): 146-154.

Watve, Y.G., Diwakar, M.P., Kadam, J.J. and Sawant, U.K. (2009). Studies on effect of different fungicides on *Colletotrichum gloesporioides* penz. causing leaf spot disease of Jatropha. *J.Pl. Dis. Sci.,* **4**(1): 120-121.

Watve, Y.G., Diwakar, M.P., Sawant, U.K. and Kadam, J.J.(2009). Studies on effect of different fungicides and Phyto-extract on *Colletotrichum gloesporiodes* Penz. Causing leaf spot of Jatropha. *J. Pt. Dis. Sci.* **4**(1): 95-98.

Wei Y.D, Shen W.Y, Dauk M, Wang F, Selvaraj G and Zou J.T. (2004) Targeted gene disruption of glycerol-3-phosphate dehydrogenase in *Colletotrichum gloeosporioides* reveals evidence that glycerol is a significant transferred nutrient from host plant to fungal pathogen. *Journal of Biological Chemistry,* **279:** 429-435.

Xu H.L (2001) Effects of a microbial inoculants and organic fertilizers on the growth, photosynthesis and yield attributes and economics of rice (*Oryzasativa*). *Crop Res.,* **31:** 1-5.

Yadav B.K and LourdrajC.A (2006) Effect of organic manures and Panchagavya spray on yield attributes and economics of rice (*Oryza sativa*). *Crop Res.,* **31:** 1-5.

Zaman, M.M., Masum, A.S.M.H., Ahmed, M.A., Salam, M.A. and Rahman, M.H. (2002). Effect of tillage and mulch on the growth and yield of ginger in the Hilly area. *J. Biological Sci.,* **2**(2): 121-123.

Zhuang Lu Yang (1992)Extraction and determination of flavonoid in ginkgo *Chinese Herbal Medicine,* **23:** 122–124.

Zitter TA. 1972. Pepper virus strain identification in Southern Florida. *Phytopathology,* 62: 800.

Index

C

D

E

F

G

Q

R

S

T

U

V

W

Y